Strange and Charmed

Science and the Contemporary Visual Arts

'Strange' and 'Charmed' were among the first names given to quarks, the smallest known particles of matter. The word 'quark' was adopted by the American physicist Murray Gell-Mann and became current in the 1960s, although it was first coined by James Joyce to represent the sound of seagulls in 'Three Quarks for Muster Mark!' in his novel *Finnegan's Wake* (1939).

I have asked the company how many of them could describe The Second Law of Thermodynamics. The response was cold: it was also negative …

C.P. SNOW, *The Two Cultures*, 1959.

… yes, disorder
is increasing in the universe

and will keep increasing until
the whole shebang becomes a place
where it is remembered
only the alert rodents swam.

JO SHAPCOTT, from 'Pavlova's Physics', 1992.

Strange and Charmed

Science and the Contemporary Visual Arts

Edited by Siân Ede

Preface by A.S. Byatt

CALOUSTE GULBENKIAN FOUNDATION, LONDON

Published by the
Calouste Gulbenkian Foundation
United Kingdom Branch
98 Portland Place
London W1N 4ET
Tel: 020 7636 5313

ISBN 0 903319 87 X

British Library Cataloguing-in-Publication Data
A catalogue record for this book is available from the British Library

Designed by Andrew Shoolbred, cover by Tim Harvey
Printed by Expression Printers Ltd, IP23 8HH

Distributed by Turnaround Publisher Services Ltd, Unit 3, Olympia Trading Estate, Coburg Road, Wood Green, London N22 6TZ. Tel: 020 8829 3000, Fax: 020 8881 5088, E-mail: orders@turnaround-uk.com

Front cover: Susan Derges, *The Observer and the Observed* (no. 6), 1991. Gelatin – silver print, 71 x 61 cm. © Susan Derges. Courtesy of the artist and Michael Hue-Williams Fine Art Limited, London. (See Chapter 7, page 123.)

Back cover: Cornelia Parker, *Meteorite Lands in Tivoli Gardens,* 1996. Courtesy of the artist and Frith Street Gallery, London.
The iron from a meteorite found in Namibia was carefully fragmented, mixed with gunpowder and exploded over the Tivoli Gardens in Copenhagen, forming a meteorite shower as part of a regular firework display which lit the night skies weekly over four months. Cornelia Parker has commented: 'The use of meteor rock as an ingredient in an artwork scandalised some Danish scientists. But in science it is common practice to break down and destroy material in order to gain new knowledge. The artist does not wring new knowledge from the material but re-interprets it, making the existing material visible in a different form.'

Contents

Preface

If you ask a city financier – or a diplomat, or a journalist, or a writer or a student – what they are just now reading, the answer is likely to be books about science. We think out ourselves and our place in the world in terms of what we know of astrophysics, or genetic research, microbiology or the study of the brain and the physiology of consciousness. This was not so 30 years ago – people thought in terms of Freud, Jung and Marx, who were all prophets who had made belief systems and issued commandments. More recondite theorists – Derrida, Foucault – have created a suspicion of thought itself, alongside a belief in the 'social construction of reality'. We have considered the idea that our language and our social circumstances speak us, not we them, and we have believed that the ideas of objectivity and reason are impossible and perhaps unnecessary. It is possible that the immediate politicisation of all ideas is diminishing with the great belief systems that – in the West – came after religions. (This is not to say that politics is unimportant. And we should remember that the United States is a predominantly Christian country, with places where Darwinian theory is considered false and evil.) We read science out of concern for our own health and environment. But I think we also read scientific books because they are the best way we now have of answering the perennial human need for understanding, contemplation of our place in the order of things, a sense of complexity and mystery, an inkling perhaps of the order of those things which are not ourselves. For the Age of Suspicion led to solipsism, to navel-gazing, to a sense that the inside of our own head was all we could know. This complacent mental misery makes no sense in the world of scientific discovery. We need to feel that there is something real out there – of which we are a part and not the whole – and science reveals it to us in its beauty and its terror and its order and its chaos, bit by fascinating bit, cell by cell, gene by gene, galaxy by galaxy. Curiosity doesn't kill cats, it saves them. It is a fundamental human drive, and the opposite of solipsism.

This fascinating book is about the relations between science and the visual arts. The artistic 'culture', as Siân Ede demonstrates, differentiates itself from the scientific culture by cherishing the individual gesture and scribble, and very often by characterising itself as the subversive, the destabilising, the contrary. As a young actress said to me enthusiastically in the 1980s, 'The theatre IS the Opposition'. Automatically? I asked. To everything? Yes, she said firmly. I sometimes think that the culture of automatic subversiveness is itself a pervasive and uniform orthodoxy, in urgent need of being questioned or subverted. We may come to see its products as the 'salon art' of our time, conformist whilst somewhere else other artists are doing something dangerous and new. Perhaps – as some of the examples in this book show – by taking science seriously, looking at it with true curiosity, making the effort to study and understand, at least in part, how scientists work and what they find.

The scientific culture is changing our moral world. This is not because it represents a belief system, although individual scientists have beliefs – socialist or social-Darwinist for instance – that do affect the way they do their science, and the arguments they put forward about the world we all live in. The passions expressed at the Darwin Seminars at the LSE show how close both Darwin's discoveries, and the opposition to Darwinist rigour, come to giving rise to belief systems. Lewis Wolpert argues steadily that it is possible – and necessary – for the work of scientists to be carried on objectively, rationally, and in the interests simply of scientific knowledge. Moral discussion of the uses of that knowledge – from genetic engineering to embryological manipulation – is the province of politicians, ethical 'experts' and the public. In literature the ethical problems are being dramatised in the 'Darwinian' novels – not only the religious crisis and loss of faith of the nineteenth century, but more subtly, the changes in our images of ourselves that come with the understanding of Darwinian arguments. I recently wrote an essay on the modern English Darwinian novel and made the surprising discovery of a constant, desperate attempt by male novelists to defend the idea of Romantic Love (not God) against the understanding of sex pheromones and the tyranny of the selfish gene.

I believe the new images and understanding we are acquiring of the biology of consciousness will slowly change the forms of works of art in many disciplines. Colin Blakemore writing (in the *Independent on Sunday*, 2 January 2000) on the new problems to be posed, or solved, by the study of the brain, says that neuroscience will 'undermine such cherished notions as spirituality, intuition and altruism – not by denying that people have them, but by providing rational accounts of them'. Some artists react to this kind of perception by fierce and immediate defence of the notions. But I cannot believe that curiosity about the science will not be more illuminating than automatic principled opposition to it. Just as understanding the complexity of genetic alterations will in the end surely produce, not naively mocking potato-headed men, or banana-tailed superlambs, but shifting metaphoric forms like Bernini's Daphne or Polke's witches and demons in caverns of poisonous and lovely pigments.

The visual arts have always had an aspect of practical science – John Gage, in his new *Colour and Meaning* (subtitled 'Art, Science and Symbolism') deplores the failure of scientific theorists of colour vision and colour mixing either to consult art historians or to study individual works of art. His account of the discovery of pigments, and the cultural effect of the difficulty of obtaining, say lapis lazuli, and its superior durability once obtained, is fascinating. Sigmar Polke uses old and discontinued poisonous pigments, lapis, orpiment, Schweinfurt green. But he does not stop there – he mixes traditional materials with iron, aluminium, potassium, manganese, zinc, barium and adds turpentine, alcohol, methanol, sealing wax and candle smoke to very corrosive lacquers. Many of his paintings are unstable, designed to change in time in unpredictable ways. Some are made with meteorite and tellurium – his work is a physical and chemical exploration of the world of Matter – though the title of the earthiest series – derived from a native American saying – is *The Spirits who Lend Help are Invisible.* The British abstract painter Patrick Heron, who worked in oils and in gouache, but also in glass and metal, always resolutely described himself as a 'materialist'. There is more to be learned, more richly, from Heron's expanses of colour, their borders and meetings and intersections, about the relative stability and instability of our colour discriminations than from a host of experiments, just as his large bright canvases fill galleries with hosts of dancing variegated after-images – all part of the work.

Heron's materials were traditional. So much more is now available to artists – cameras and optic fibres, radioactive isotopes and x-ray films, microscopes, telescopes and cellular materials. This book gives some splendid examples of artists working with flesh and blood, embryos and sperm. There is a sense in which artists have to fight to make art and *to know that it is art* in a world full of seductive, delightful, intriguing and brilliantly crafted technological visions, from the juxtapositions and framed geometries of commercials to the beauty of time-lapse films of the growth of plants and the movements of air, earth and water, which, like early photographs of race-horses, provide new scientific information as well as aesthetic delight. Video art and installations often have a hand-held, amateur look which is more boring than moving to an audience accustomed to the stimulation of the world of the web and the light show and the television. What (besides being made by an individual and *labelled* as art) makes art art? All my life I have urgently been asking myself why human beings make works of art at all – there is no obvious moral or Darwinian reason why they should. The answer, often enough, seems to be, for the same reason that they make metaphors. I don't understand that either – except that metaphor-making is a fundamental part of the way the human mind makes connections. And art explores connections like those in ways very different from science's orderings – even though scientists are aided by flashes of intuition.

There are subtle and elaborate explorations of connection-making in the chapters of this book, from the labyrinthine fictive or alternative taxonomies of Chapter 5 to the bodily metaphors of Helen Chadwick and Cornelia Parker. Connection-making itself in the age of neuroscience can be imaged, can almost be *felt*, as a subtle movement and flashing of the neurons and synapses of the brain – creating and recreating in memory and imagination representations of things meeting and parting, arranging themselves and dissolving. Scientists, without embarrassment, use the words 'beautiful' and 'elegant' about mathematical solutions and the structures of cells and stardust. They tend to expect artists to be interested in beauty, and to be surprised to find that we have a stock response that 'beautiful' things *must be* sentimental or facile. The culture of subversion, and suspicious solipsism make it hard to admit that a need for beauty has driven human beings for as long as a need to make connections. And the idea of beauty is connected to Gombrich's idea, so persuasively explored in Mike Page's chapter, of knowing when something is 'just right'. We see a Patrick Heron, or a Polke scribble, and know without knowing how, that it is 'just right', that moving a part of the form would destroy it. Coleridge called this just-rightness 'cohaerence', using the Latin form to display its ancient meaning, of clinging, or holding, together. Artists and scientists both recognise cohaerence. Artists, as Siân Ede argues, often bring an alien sense of order and connection to a scientific object or concept, that can reveal new things in it. Again, it is like metaphors. It's like the difference between the sentences of ordinary language philosophers – 'There is an apple on that table' and

This my hand will rather
The multitudinous seas incarnadine
Making the green one red.

You can argue for days about the meaning and technique of that, but its effect of complex and multiple interactions of language and feelings is instantaneous. It makes the hair stand up on my neck. There is a scientific explanation for that, I'm sure. De Quincey called his great and interlinked dream metaphors 'involutes'. The immediate, and subtle, and complex involutes of art can reveal new connections in both language and the material world. It increases wisdom and understanding – and pleasure. I'm with the scientists, not the automatic opposition, in finding a need for beauty human and desirable – whatever strange new forms it may make.

A.S. Byatt

Foreword

There is no natural reason to link science and the visual arts. There is also no reason why not. In a post-modern culture any two unrelated subjects can be held up against each other simply for intellectual diversion. But what's the point? This book will need to engage with intellectual ideas because a study of art and science requires abstract thinking, but it will be for practical ends. Produced by an arts funding body, it tries to address the real world, even while the 'real world' will be shown to have very different meanings for artists and scientists.

The real world for the Gulbenkian Foundation's Arts Programme is concerned with the spending of a limited amount of money on art and artists. Since that money is scarce, it is always pertinent to question why we choose to support one thing above another. The Foundation creates new funding programmes, observes the consequences and sometimes publishes books to examine the field, with deliberations about practical facilitation as well as theoretical speculation, especially in the context of the social, political and funding climate. The Foundation has been a modest but important funder of the contemporary arts in the UK for over 40 years. It is generally acknowledged for its part in establishing contemporary dance in the UK in the 1970s. Its 1982 publication *The Arts in Schools* has had a significant influence on public education debates. In the visual arts, its 1985 publication *The Economic Situation of the Visual Artist* was the first piece of detailed research into an underworld of ingenuity and deprivation.

Our greatest challenge has been to recognise genuine innovation and, in so doing, to persuade ourselves that we have played a part in identifying, if not inventing, new movements. If the current surge of interest in 'Art and Science' is anything to go by, it may well turn out to be a new and significant area of activity. This book's simple intention is to show how a knowledge and understanding of science can influence the practice of contemporary art. It is principally for artists, arts facilitators, funders and decision-makers, arts educators and students. It should also interest scientists and some of the wider public who may gain a perception of how, conversely, a knowledge and understanding of contemporary art can influence the practice of science.

The book glances over a huge terrain, taking into its scope many aspects of the arts and sciences with passing references to philosophy, cultural theory and politics. Every statement uttered could lead to a finely branched network of further thoughts and arguments. If we have succeeded we will have sketched out the broad picture but left it up to readers to pursue particular areas in more detail, giving them licence to adjust and contradict our broad configurations as they do so. This debate – the confrontation between art and science – is a living one and we know that we have not made definitive statements. Scientific discovery leaps ahead fast and cultural discussions take unexpected turns. It is our hope that readers are able to make a virtue of our unavoidable omissions and find them enough of a stimulus to pursue their own trajectories.

• • • •

We are particularly grateful to the members of the Steering Group of arts and science practitioners who provided much of the inspiration for this book, and with whom, as a group and individually, there were many stimulating discussions. They are: Marjorie Allthorpe-Guyton, Ken Arnold, Frances M. Ashcroft, Paul Bonaventura, Richard Bright, Daniel Brine, A.S. Byatt, Andrea Duncan, Stephen Farthing, Christopher Isham, Melanie Leech, Gillian Thomas (Chair) and Nicola Triscott.

We are also very grateful to Frances M. Ashcroft, Paul Bonaventura, Christopher Isham, Christine Kenyon Jones, Tom Sorell and Jenny Uglow, who acted as specialist advisers and readers. Many other people, from both the arts and sciences, have helped with research, meetings, correspondence and checking, and are acknowledged on page 189.

Particular thanks should go to Christine Darby and Felicity Luard at the Gulbenkian Foundation who have made an invaluable contribution, from the original conception right through to the production of the book, and also to Jane Barry, Catherine Goddard, Tim Harvey, Sue Rose, Andrew Shoolbred and Sue Ullersperger.

Siân Ede
Assistant Director (Arts)
Calouste Gulbenkian Foundation

1

Bizarre Consequences: Introduction and Background

Siân Ede

... believe me, this world
is a place of bizarre consequences
where matter can appear
out of nothing and where

the light of stars is ancient
history when it gets here:
we can never understand
what we're living through at the time.

JO SHAPCOTT, from 'Pavlova's Physics'.[1]

Why has an arts funding body turned its attention to science?

Because not a day goes by without a major news announcement which begins: 'Scientists have discovered that ...'
Because the revelations of science are curious and beguiling ...
Because science provides transformative world-views ...
Because the political and practical implications of science require urgent consideration ...
Because some artists passionately want to work with science, its ideas, its new materials, its opportunities for engagement in a different world of work ...
Because an examination of science forces us to view art from new perspectives.

'A knowledge of science sets free the idea that the universe is not about you and stimulates you to make use of your time here,' says novelist Ian McEwan. We hear about quantum theory, chaos, evolution, the nature of consciousness, black holes and baby universes, nuclear waste, chemical warfare, genetic engineering, space exploration, global warming. Science books are regularly on bestseller lists. Literary journals contain an increasing number of reviews and discussions on science alongside those on the arts or humanities. *New*

Mark Wallinger's sculpture of Jesus is recognisably life-size, although, a lone slight figure on a high plinth in Trafalgar Square, London, he is easily dwarfed by the grand architectural proportions of the site. Consider his scale, however, set against the spatial and temporal vastness of the universe. Conversely, imagine the minuteness of the elements of which the human body is composed.

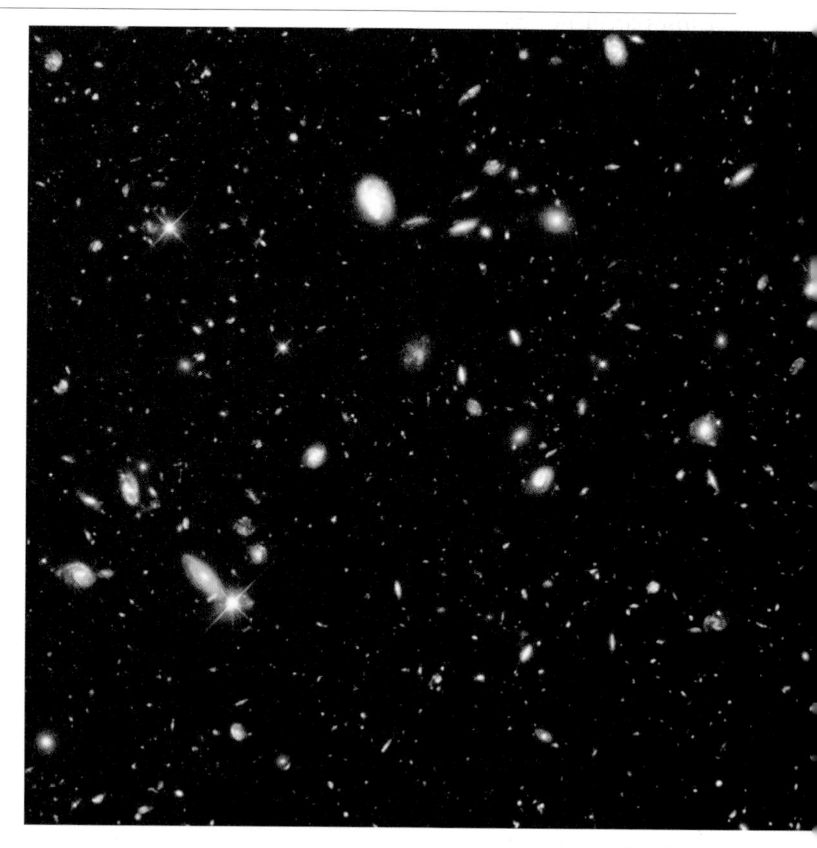

Mark Wallinger, *Ecce Homo*, 1999. White marblised resin, barbed wire and gold leaves, life-size. © Mark Wallinger, courtesy of Anthony Reynolds Gallery, London.
Photo: Liam Bailey.

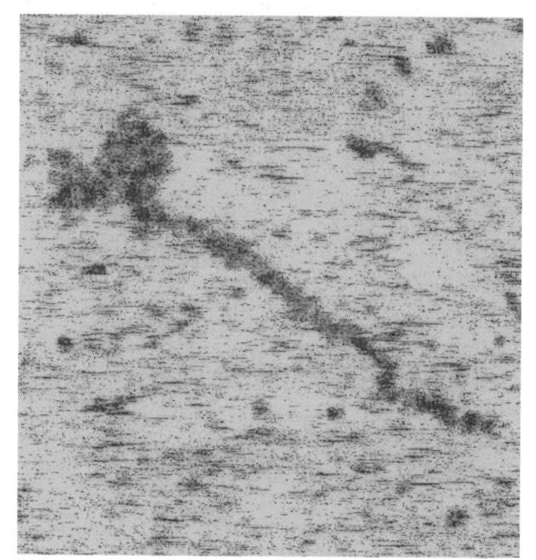

Scanning Probe image of one of the smallest molecules to be captured on film, a myosin molecule, muscle protein. The tail is 155×10^{-9} metres. Courtesy of Peter Hallett/Wellcome Trust Medical Photographic Library, London.

The Hubble Deep Field is the deepest view of the universe ever obtained. This image was recorded by the Hubble Space Telescope over a period of 10 consecutive days in December 1995. The very faintest galaxies visible are some four billion times fainter than the faintest objects that can be seen by the human eye. The most distant galaxies detected in the Hubble Deep Field, which are visible in the image as small red dots, are some 15 billion light years away, at the very edge of the observable universe. Light from these very distant galaxies was emitted soon after the Big Bang.
The Hubble Deep Field. © Astrographics Publishing, Smithtown, NY.

Scientist's circulation has increased by almost 25 per cent since 1993. The London School of Economics has turned to Darwinian paradigms and new science-inspired theories are emerging in politics and sociology. Even the talks programme at the Institute of Contemporary Arts is as likely to feature neuroscience or genetics as Lacan, Baudrillard or cyborgs.

We are turning to science, says the writer A.S. Byatt, 'because the many new popular science books make it appear accessible', although she is quick to point out that many pursue 'a desire to believe in knowledge and order from a position of clerkly scepticism'. There is, however, a real indication that we are looking for something more than a challenge to the intellect or a novelty of setting. 'The big belief systems of the recent past – Marxism, Freudianism – are no longer believed, even if they live on,' she writes. 'The strenuous refusals of order and meaning of Foucault and his school are exhausting and circular. And beyond that we have local belief systems, good ones – anti-racism, anti-sexism – based on moral passion but also given to closed circularities of argument.'[2] When we observe scientists operating in dimensions beyond the familiar certainties of time and place and scale, what new meanings do we perceive for our lives and ways of thinking about ourselves? When we learn about the thought experiments and theoretical models scientists engage with, their appropriation of human metaphor for non-human occurrences, their descriptions of the unnatural behaviour of phenomena and events which operate beyond our narrowly confined human existence, how is our world view altered?

Michael Frayn's play *Copenhagen* is an ingenious speculation about what may have occurred during a wartime meeting between the Danish quantum physicist Niels Bohr and his former colleague Werner Heisenberg, then working for the Nazis. Both were involved in research to split the atom. In presenting a number of possible versions of their encounter Frayn draws on quantum theory, particularly the notions of *uncertainty* and *potential*, using them specifically, not in their usual sense but as conceits to illustrate how we continually remake inexact versions of the truth, the act of measurement itself determining the status of the event.

> 'To understand how people see you we have to treat you not just as a particle, but as a wave. I have to use not only your particle mechanics, I have to use the Schroedinger wave function …,' says Bohr.
>
> Heisenberg continues the conceit: 'Exactly where you go as you ramble around is of course completely determined by your genes and your upbringing and the earth's magnetic field and the gravitational pull of the moon. But it's also determined by your own entirely inscrutable whims from one moment to the next. So we can't completely understand your behaviour without seeing it both ways at once, and that's impossible, because the two ways are mutually incompatible. So your extraordinary peregrinations are not fully objective aspects of the universe.'[3]

Such dialogue distils the old philosophical discussion about fate and free will, engaging paradigms from quantum physics for dramatic purpose and in doing so throwing fresh meaning on human motivation. But it takes a precise sense of irony not to make the scientific conceits too clumsy and literal. If only physics were so easy to understand. Different vocabularies are enticing to any artist, writer or theorist who is looking for fresh ways of attracting attention but the original concepts are often extremely difficult to understand properly, especially when they are normally exchanged through the currency of mathematics. Familiar words have different meanings. Images and models represent far more than meets the eye. Scientific concepts may provide more than new bottles for old wine.

Paradigms gained

Contemporary science continues to produce new metaphors and paradigms for viewing the world. Some will remain self-contained, relating only to the discrete and difficult area of science from which they have developed. But others are being claimed by scientists as important indicators of new frameworks for our knowledge about the world, literally and metaphysically. The arts community, meanwhile, remains preoccupied with its own belief systems and one of the most striking distinctions between the two cultures is that they appear to occupy two separate churches with entirely separate doctrines. Let us briefly look at some new scientific concepts.

'Perhaps the universe is really a frolic of primal information,' says the physicist Paul Davies, 'and material objects a complex secondary manifestation.'[4] Our corporeal existence, the neo-Darwinists claim, results purely from random mechanisms which replicate information, selected 'naturally' on the basis of their survival alone.[5] It is 'all in our genes'. Not everyone agrees. The context in which genes operate – within the single cell, within the organism, within the external environment – is just as important in determining corporeal existence and behaviour, to say nothing of our mental processes.[6] But neo-Darwinism prevails for many. We are still hunter-gatherers by nature and evolutionary psychology explains all our motivations.[7] Everything 'evolves'. Perhaps even our thoughts, ideas, skills, fashions and stories are discrete packets of information, mental self-replicating genes or 'memes' (a term invented by Richard Dawkins) passed on like a virus from person to person. Our cultural standpoints, belief systems, religions and political ideologies are 'memeplexes' which evolve through imitation and copying, propagating (presumably if 'good'), infecting (if 'bad'). Will the paradigm of 'memes' catch on or is it simply an elaborate construct?[8] (If that is the case, aren't contemporary artists by their very nature 'meme-breakers', or at any rate, 'meme-initiators', because they question, alter, invent, insist on individual dissension?)

Our brains may be simply sophisticated artificial intelligence processors. On the other hand, perhaps a moment of consciousness can be tracked down, according to the mathematician Roger Penrose, to fluctuations of electrical activity in the neurons of the brain. Perhaps the flickering of that quantum event can actually be observed by Buddhist meditators who have developed over many years the facility to will themselves into higher states of consciousness? Strict brain scientists would not agree. However, the word 'consciousness' has been appropriated by a whole range of scientists, as it has by cultural critics, theologians and philosophers, who apply to it an assortment of different meanings. The mystery of mind remains. How can mere systems of information-processing result in our singular experience of a rich inner life?[9]

The concept of digital information-processing may be no more or less than the metaphor of the age. But it begs questions about how the capture, replication and application of information affects the way this information is interpreted. How do we distinguish between the medium and the message? Both digital science and digital art exploit a dislocation of our sense of the 'virtual' and 'real', the artificially intelligent and the imagined. The old debates are revived: mind or matter, fate or free will, determinism or choice, nature or nurture; but the choices may not be so stark. Biologists have become increasingly materialist and deterministic but some are still susceptible to a belief in ideas which may be simply constructs (like the idea of 'memes'). Some physicists have been drawn to Jung's concept of synchronicity or 'meaningful coincidence', the possible a-causal connection between psychic states and objective events, and have sought to apply it to notions of discontinuity, non-locality and entanglement in quantum mechanics.[10] When two particles interact they become inextricably linked forever even if they travel to opposite ends of the universe. This relates to the idea that all particles originated in the same dense singularity before the Big Bang, when all was one and one was all. But despite appearances, physics is not becoming another expression of New-Ageism. Many physicists look to a Grand Unified Theory which will rationally tie up all the explanations.

What is the nature of reality? It is fascinating to observe that scientists deal not simply with matter but, as the 'natural philosophers' they originally were, with conceptual thinking, with the contemplation of deep meanings and with epistemological questions. The arts constituencies have their own discourses but they should not ignore the emerging paradigms of science, for these are already becoming part of the currency of the wider world and deserve to be taken seriously.

Experiments in art and science

It would be hard to find examples in the modern period, and beyond, of occasions when artists have not in some way become involved in science, even if unconsciously or tangentially. However, it is not the purpose of this book to re-examine artworks with a new vision informed by science. Fractals and emergent systems can appear wherever you look for them, if that is your fancy, but it would be a dull art critic who sought to analyse art in purely physical terms. Art requires deeper analysis, not a superficial appropriation of scientific terms because they happen to be all the rage. Art historian Martin Kemp (who read Natural Sciences at Cambridge) has written perceptively and eloquently about the complexities concerning the inter-relationships between science and art and has been highly praised by many scientists for his column in the prestigious science journal *Nature*.[11] The historian of science Arthur I. Miller has explored the coincidental connections between modern art and modern physics and shown how the *zeitgeist* informs the way we express our meanings at particular periods in history.[12] In this far less academic book we will be considering the works of particular contemporary artists who consciously address science and we will also examine what has become a new phenomenon, the setting up of art-science collaborations to make new work which reflects the interests of both disciplines.

The pioneering work of British art-science organisations – London-based Arts Catalyst, The Laboratory at the Ruskin School of Drawing and Fine Art at the University of Oxford, Bristol-based Interalia, the Medicine, Society and History Division of the Wellcome Trust, and the Gulbenkian Foundation's own *Two Cultures* programme – will be described in Chapter 3. What is worth noting here is the way in which these enterprises have reflected an increasing mutual curiosity, if not an appetite for confrontation, from the constituents of both worlds and from the public beyond them. The Institute of Contemporary Arts has collaborated with the Royal Institution, for example, in holding a series of science lectures, for which it has become increasingly hard to get tickets. The Wellcome Trust held two packed conferences at the culmination of its 1997 and 1999 Sciart programmes, in which the successful collaborative projects were shown as lecture demonstrations – a format familiar to the science world. The Arts Catalyst held a conference in 1998, 'The Eye of the Storm', which brought to arts audiences some of the world's leading scientists who came to talk about their research and debate a very wide range of issues with artists. In these events it was a revelation to those of us in the arts and humanities to discover that when it came to explaining their work and its implications, scientists were performers of world class. The substance of their talks, moreover, about the nature of consciousness, artificial intelligence and genetics was often riveting for the lay audience. Many also speculated about the deeper abstract meanings they perceived in their areas of research. As if taken by surprise, it is fair to say, most of the artists did not communicate half as well and this was troubling. Visual artists are not, of course, expected to use language fluently (although some can) for it is not their primary medium, but artists from any art form might find it hard to compete with an intellectual community for whom the language of rational explanation is part and parcel of the disciplinary approach to the work. Gifted arts speakers might use evocative narratives, anecdotes or pictures, but their assertions cannot stand up to an agreed analytical scrutiny and it may well indicate that this difference in modes of communication is a fundamental factor in the divide between the two cultures. Certainly there were members of both constituencies who went away mystified or cross. Why, then, are they still drawn to each other?

Visual controversies

Encounters with science and scientists have been open to artists from all art forms but most surprising is the fact that they have attracted primarily visual artists. It is easy to see similarities between scientists and musicians, with their fastidious desire for precision and agreement, their practice of working in groups. Theatre practitioners and literary writers, too, are able to work with conceptual, verbalised ideas. But why are contemporary *visual* artists, used to communicating in perversely abstract ways and to expressing a resolutely individual vision, drawn to science, this most rational and reductionist of activities, one whose *raison d'être* is its materialism, its strictly agreed protocols, its faith in facts? What does this mean for the visual arts world as a whole?

Visual art is everywhere in the wider world. There has never been such an explosion, such a variety of visual imagery in all its brilliance and provocativeness – in advertising, on TV, in the streets, home, school and supermarket. Visual artists have with great impact colonised other art-forms – film, theatre, pop music and pop video, dance, craft, design, fashion. But what about art for art's sake? From some perspectives it is flourishing. Galleries all over the country host an ever-changing range of events and for some exhibitions audiences are queuing round the block. There is a surge of commissions everywhere for art in public places, installations and site-specific live events. This may all add to life's pleasures but, even so, contemporary art is not well understood in the public domain. In a public culture which puts great emphasis on measurable 'objectives' and 'quality', is contemporary art a vocation without a function?

Function is a word which a few years ago would have sounded scandalously utilitarian, but many in the art world are themselves posing the question. 'There is nothing wrong with art being popular,' writes critic Robert Garnett, 'or of artists being opportunists, careerists or rampant hedonists, except for the fact that none of this can add up to a significant practice.'[13] If visual images are so effective in other domains does art still have a social role in itself? Undoubtedly it used to, but arts practitioners in general, including those from other contemporary art-forms who are used to innovative work and who need no persuading of the potential power of the arts in general, are confused and sceptical about some of the contemporary visual work they see. It is not this book's purpose to enter far into this debate and indeed it would be impossible to write if we were not committed to a fundamental belief in art's liberating powers. There is much excellent work, examples of which will be described with enthusiasm. However, in pitting art against science its potential for pointless absurdity appears to be highlighted and it is dismaying to hear expressions of incomprehension. In a controversial statement, curator Paul Bonaventura has expressed particular concern about public perceptions that a lot of contemporary art is self-indulgent. Addressing the question of creating a new curriculum for the education of young artists, Bonaventura writes:

> ... to be persuasive in their pronouncements and for their questioning strategies to be of more than local interest, artists need to operate in the margins from positions of knowledge, not as refugees from the intellectual high ground. It is a sad irony that, in many cases, some practitioners parade their ignorance about their particular interest as a badge of credibility. As a consequence their activities are treated as irrelevant by important opinion-formers.[14]

This statement cannot be regarded as true of all artists practising everywhere and the criticism is reserved for the more shallow and narcissistic of artworks, but it touches a nerve, especially if the rest of the art-world is perceived to be tarred with the same brush. If art, as the expression of an independent or individual vision, is to be regarded as a thing unto itself and disconnected from serious public life, this is worrying. The question is never raised whether science is perceived to occupy the intellectual high ground. It does, and at a time when the pronouncements of science have such portent, we need artists to participate and respond in their unique and original way, from a position of authority and respect. Rational discourse is not the only way we make sense of the world. 'Humans make meaning, for themselves and for others, of which they have no direct or immediate awareness. People make more meanings than they know what to do with.'[15] Artists are well placed to express alternative meanings and they should be heeded. They can be leaders and initiators too and their work should speak authoritatively and move us to contemplation whether it is seen in public exhibitions, as work in progress or through collaborative ventures in other work arenas.

This is a difficult agenda to set because art does not communicate straightforwardly. Indeed, art which baldly engages with 'issues' is usually boring art. It is as if Shakespeare had been asked to write a play merely to address the problems of geriatric care. Fortunately he came up with *King Lear*. A work of art has many resonances and its engagement with an intellectual issue may be reflected only subtly or tangentially. Viewers, the public, need to know how and where to look and this can only happen when they get used to seeing work which successfully makes them feel – and consequently, think. But perhaps they need to learn to understand the act of seeing in itself so they can recognise its value as a distinctive form of knowledge.

It is possible to identify two main strands running through twentieth-century art: a non-representational, abstract and subliminal way of seeing and making work – from Picasso, through to abstract expressionism and minimalism – and the conceptual inheritance popularly invoked in the spirit of Marcel Duchamp, where ideas predominate – usually consciously and intellectually distorted and subverted. Scientific studies of how we actually see the world show that we simultaneously perceive abstract patterns and form concepts about them; the one activity informing the other. We take in signals and interpret them. And so too with the discipline of science itself. Concepts are tested through pattern recognition – visual, behavioural, mathematical. In turn patterns suggest new concepts. Reason and logic, the detective work implied in 'the scientific method', play an important part but it comes as a surprise to learn how much scientists need to *see* or visualise ideas in order to understand.

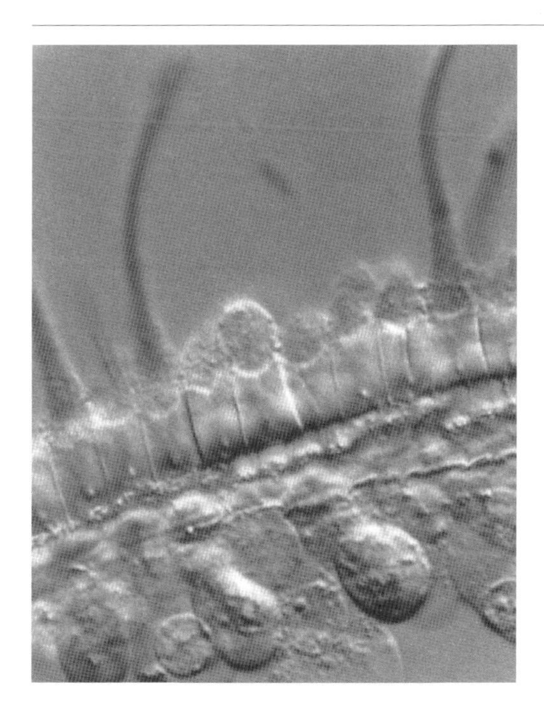

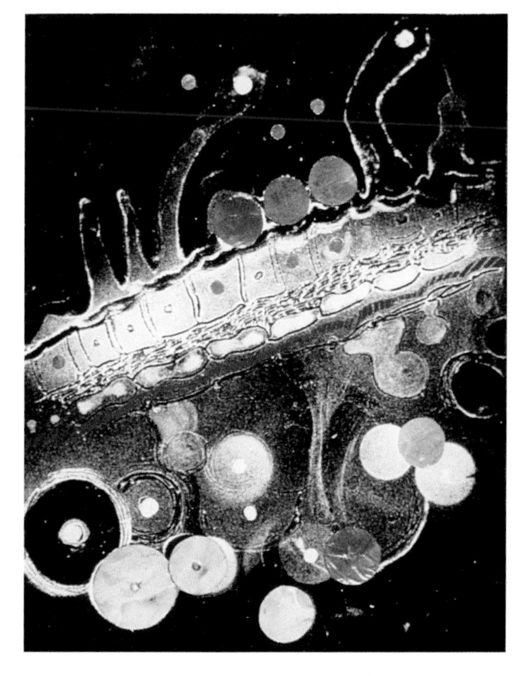

A scientific image frozen in time: fragment of the organ of Corti, the sensory structure in mammalian ears that converts sound into nerve signals. High-resolution light microscope. Courtesy of Dr M.C. Holley, Department of Physiology, University of Bristol. Photo: M.C. Holley.

The same image 'fixed' as an artwork. Sandra Macqueen, *Fragment of the organ of Corti*, 1994. Vitreous enamel on a copper background, 282 x 211 cm. Courtesy of the artist and the Wellcome Trust, London.

Moreover, current scientific research, and simultaneously the most recent expression in contemporary art, both depend on an increasing conviction that nothing can be fixed in time or place or isolated from its environment. The laboratory specimen, like the frozen painting or sculpture, is an artificial construct. Activity within the individual living cell, neurons firing in the brain, the inter-relationships between subatomic particles, the behaviour of weather and waves and insects and people are all dynamic systems in a perpetual state of flux. Both artists and scientists are aware of the impossibility of capturing the instant – the precise point when an ion channel opens, when a neuron fires; the imperceptible change of light, the gradual dying of leaves; the processes of thinking, meeting, turning away. While artists still make fixed works in traditional ways, there is an overwhelming urge to convey the fragility of the moment and the very act of change itself. It is no surprise that video, film, digitised image-making, performance, live art, superfictions and installations are increasingly used as expressive media. 'The universe is a single, unfolding self-organising event, something more animal than machine,' states architect and cultural observer Charles Jencks, citing the 'complexity or non-linear architecture' of Frank Gehry's crumpled titanium-clad Guggenheim Museum in Bilbao, 'something radically interconnected and creative, that jumps suddenly to higher levels of organisation and delights us as it does.'[16] The artists who are now turning to science may be doing so because they are engaged by its new ideas and concepts and, being always drawn to new ways of seeing, are naturally drawn to the changing patterns in living systems.

At last, then, artistic expression seems to be moving away from the existential angst and self-defeating ironies born of the political and psychoanalytical preoccupations of the recent past, towards a tougher, scientifically and technologically informed curiosity, both clinical and passionate at the same time. But in turning to science, artists force a human perspective into those chilly disorientating environments. If scientists try to abnegate the human element in order to observe nature with the utmost objectivity, artists, in contrast, place human concerns and imaginative perspectives at the centre of their work. Facts are interesting but surely only ever in relation to ourselves and how we live our lives? Tolstoy's assertion still partly obtains: 'Science is meaningless because it gives no answer to our question, the only important question for us, What shall we do and how shall we live?' Contemporary artists do not go on to dismiss science but they continue to ask the question.

It must not be supposed that artists, long attracted by the notion of alchemy, simply regard science as another way of viewing the mysterious magic of the universe. The new work which is being made is hard-edged, confrontational, employing reason and intellectual understanding, often of very difficult concepts, in order to draw out new meanings for humankind. Artists will always want to challenge consensual opinions and world-views dependent on the seemingly impartial methodologies of science but they can

'Complexity or non-linear architecture', model of a section of the Guggenheim Museum Bilbao, designed by architect Frank O. Gehry and opened in 1997.
CORBIS/Yann Arthus- Bertrand.

also learn from them and consequently begin to reappraise the status of their work. 'What is interesting for me, is that if you work with another discipline, you start taking on the terminology of that discipline,' writes artist Simon Callery, who, with photographer Andrew Watson, worked with archaeologists from the Institute of Archaeology at Oxford at the Segsbury Iron Age hill fort. 'You look at your own work with that terminology, and the whole thing becomes fresh again. And that's what I hope they get from working with me … in a straightforward way, I want to communicate that art-making requires serious thinking.'[17]

Strange and charmed

We hope in this book to investigate the nature of the confluences between art and science, using the languages both of rational explanation and of persuasive narrative and pictures. The material of science is worthy of study for its own sake. Science *is* 'unnatural' and potentially threatening, but it is also quite astonishing.[18] Once we find out a little, our world view, indeed, our perspective on life in general, is forever changed: how the human brain works; how vast the

Murray Robertson, one of a series of 'Periodic Landscapes' from a visual arts and science collaborative project called *109: A Visual Interpretation of the Table of the Elements*, 1997. Reproduced by permission of the artist and the Royal Society of Chemistry.

The Periodic Landscapes were created digitally. 2D graphs of the properties of the various elements were transposed into 3D bar charts from which the values were rendered as greyscale height maps. These maps were then imported into 3D modelling software and a fractal landscape generated. At this point the visual process took over and the models were sculpted to achieve a sense of the general trends or patterns prevalent within the values chosen.

universe is and how it may be only one among a possible 'froth' of universes; how the mitochondria, the powerhouses in our cells, operate, about the accidental beauty of elephants and inevitably, too, about the relentless decline of species across the planet. The distortions of time and place and scale glimpsed through an acquaintance with science suggest an imagined landscape of epic proportions. Inevitably we become less preoccupied with the minutiae of our lives and we are forced to reconsider our place in the universe.

Throughout this book we are rather elliptical in our use of the terms 'Art' and 'Science', as if they were each undifferentiated entities. The 'visual arts' encompass a whole range of practice which includes painting, sculpture, mixed media, photography, craft, design, architecture, art film- and video-making, installation, live art, performance art and digital art. (This is not to take into account other branches of the arts which impinge on visual arts practice, such as the many kinds of music, drama, literature, dance and feature film.) And, of course, the work of no two individual artists is alike. The term 'science' encompasses such very different practices as physics (theoretical and experimental, for example), the medical sciences, neuroscience, biochemistry, molecular biology, genetics, the environmental sciences, chemistry and geology, and also includes those which though more open to interpretation still rely on scientific methodology, such as anthropology, archaeology, psychology and many more. Technology and engineering are different from science itself, though normally associated with it. Lewis Wolpert's distinction is that 'science produces the ideas whereas technology results in the production of usable objects'.[19] Among themselves, groups of practitioners within each scientific discipline will often have strong disagreements – conceptual, interpretative and also ethical. Nevertheless, all those disciplines which apply 'the scientific method' to their routine practice are fundamentally different from the arts.

The following chapter, entitled 'The Scientist's Mind: The Artist's Temperament', explores how the two cultures have developed to reflect quite separate ways of looking at the world. The next one, 'Many Complex Collisions', however, shows how they may usefully come together at times and enrich each other's practice. Ken Arnold in 'Between Explanation and Inspiration: Images in Science' examines the visual practice and product of science and its influence on art. In a chapter entitled 'Us and Them, This and That, Here and There, Now and Then', Martin Kemp and Deborah Schultz question the rudimentary bases of orthodox classification systems in order to present the work of artists who have imaginatively subverted taxonomic principles, only to make original reclassifications. Mike Page in 'Creativity and the Making of Contemporary Art' explores the mind processes of both artists and viewers, with particular reference to Richard Wentworth's 1998–9 exhibition *Thinking Aloud.*

It would be impossible to investigate each of the many disciplines that belong within the overarching reach of contemporary science so we have chosen to explore some aspects related to two of its major pillars, physics and biology. In 'Uncertain Entanglements' Richard Bright clarifies some of the basic concepts of theoretical physics, conundrums which lend themselves to metaphysical interpretation but which are often far beyond the understanding of non-physicists and suggests that artists need to be wary and avoid seizing too easily on words and ideas which do not have ordinary meanings. In her chapter 'Inside – Outside – Permutation', however, Andrea Duncan investigates how certain contemporary artists have grasped the opportunity to comment on the human condition by presenting the body in arresting new ways as a result of their delving into new biomedical research and imaging techniques. And in the final chapter, 'Fission or Fusion', we examine the work of artists who engage with some of the complex moral and political questions associated with scientific practice. The book ends with a brief list of science resources, and includes arts organisations involved with art and science collaborations.

This investigation will not turn artists into scientists nor scientists into artists. The two cultures are distinct, although all of us would benefit from developing our understanding of both, consequently acquiring a broader view of how the 'real world' operates. The leading experimental psychologist Richard Gregory has written:

> Experiencing and thinking are different and may not go well together. Admiring a sunset and at the same time thinking out why the sky is red, evoke different mind-brain processes, which few of us like to switch on simultaneously. Listening to a symphony on the radio, one may be puzzled that the different sounds of all those instruments can be conveyed by the vibrations of the single cone of the speaker. One may not want to bother with this while listening to the music, but surely it adds to life at least to see it as an intriguing question. Explanations can make us see pictures, and everything else for that matter, in a very different light. And why not? Is it not fun to switch mental views? Surely the trick is to live in at least two cultures, though not necessarily at the same time.[20]

It remains to be seen whether simultaneous translation is possible.

James Turrell makes works which challenge the ways in which we construct reality through observation and interpretation, creating disorientations through manipulating our perceptions of space and light. In his walk-in installations he creates environments where the viewer becomes unable to distinguish between objective and imaginative ways of seeing. In *Roden Crater* he is turning a large volcanic cinder cone in the Arizona desert into a complex of viewing chambers from which to experience the varying light from the sun, moon and stars, enabling viewers to create new individual psychological spaces. 'My works are about light in the sense that light is present and there. The work is made of light. It's not about light or a record of it, but it is light. Light is not so much something that reveals, as it is itself the revelation.' (James Turrell, quoted in *Occluded Front*, exhibition catalogue. Los Angeles, The Museum of Contemporary Art, 1985.)

James Turrell, *Roden Crater*, 1986. Drawing, original crater shape for full-scale installation. Photograph, wash, pencil and marker pen on mylar. Courtesy of Michael Hue-Williams Fine Art Limited, London.

Notes and references

1 Jo Shapcott, 'Pavlova's Physics', from *Phrase Book* (Oxford, Oxford University Press, 1992).

2 A.S. Byatt, 'Belief in the Jungle of Ideas', in *The Guardian*, 29 August 1998, a review of *Consillience* by Edward O. Wilson (New York, Little, Brown, 1998).

3 *Copenhagen* opened at the Royal National Theatre's small Cottesloe theatre on 28 May 1998 for what was assumed would be a coterie intellectual audience, but sold out rapidly. It subsequently transferred to the West End, toured nationally and to Connecticut and Malta, and won three British theatre awards, including Best Play of the Year for the *Evening Standard* and for the Critics' Circle Drama Awards. It also won France's top theatre award. Michael Frayn, *Copenhagen* (London, Methuen Drama, 1998), pp. 71–2.

4 Paul Davies, 'Bit before it?', editorial in *New Scientist*, 30 January 1999; related feature, 'I is the Law', pp. 24–8.

5 For a first look at neo-Darwinism see Richard Dawkins, *The Selfish Gene*, 2nd edn (Oxford, Oxford University Press, 1989), and *The Blind Watchmaker* (London, Penguin Books, 1986); Matt Ridley, *The Origins of Virtue* (London, Viking, 1996); Daniel Dennett, *Darwin's Dangerous Idea* (London, Allen Lane, The Penguin Press, 1995; New York, Simon and Schuster, 1995).

6 Opposing strict neo-Darwinism see Steven Rose, R.C. Lewontin and Leon J. Kamin, *Not in our Genes* (London, Penguin Books, 1984); Stephen Jay Gould, *Leonardo's Mountain of Clams and the Diet of Worms* (London, Jonathan Cape, 1999); Mae-Wan Ho, *Genetic Engineering: Dream or nightmare?* (Bath, Gateway Books, 1998).

7 For evolutionary theory see series *Darwinism Today* (London, Weidenfeld and Nicolson, 1999); also Steven Pinker, *How the Mind Works* (London, Allen Lane, The Penguin Press, 1997). For a light-hearted spoof of the worst excesses of evolutionary psychology, V.S. Ramachandran, 'Why Do Gentlemen Prefer Blondes', appendix to V.S. Ramachandran and Sandra Blakeslee, *Phantoms in the Brain* (London, Fourth Estate, 1998), pp. 289–91.

8 Memes – see Richard Dawkins, 'The Balloon in the Mind', in *Unweaving the Rainbow* (London, Allen Lane, The Penguin Press, 1998), and Susan Blackmore, *The Meme Machine* (Oxford, Oxford University Press, 1999).

9 For an introduction to British consciousness studies see Steven Rose, ed., *From Brains to Consciousness* (London, Allen Lane, The Penguin Press, 1998); Susan Greenfield, *The Human Brain: A guided tour* (London, Weidenfeld and Nicolson, 1997); records of the annual Tucson conferences on consciousness published in *Journal of Consciousness Studies: Controversies in science and the humanities*, ed. J.A. Goguen and Robert K.C. Forman (Imprint Academia; ISSN 1355 8250); see also R.L. Gregory, *Eye and Brain* (Oxford, Oxford University Press, 1970); Ramachandran and Blakeslee (1998), see note 7; Oliver Sachs, *The Man who Mistook his Wife for a Hat* (London, Gerald Duckworth, 1985).

10 The physicist and Nobel prizewinner Wolfgang Pauli (1900–1958) was one of several scientists interested in Jung's ideas and sought to add synchronicity to the notions of time, space and causality in classical physics to create a new theoretical synthesis. The idea of combining quantum theory with a Jungian blend of mysticism, alchemy, mythology and psychology is still very appealing to some New Age thinkers seeking a holistic approach to explain the relationship between the cosmos and the spiritual world.

11 Martin Kemp, *The Science of Art: Optical themes in Western art from Brunelleschi to Seurat* (New Haven and London, Yale University Press, 1990), series in *Nature* published as *Structural Intuitions: The 'Nature' book of art and science* (Oxford, Oxford University Press, 2000), and *From a Different Perspective: Visual angles on art and science from the Renaissance to now* (New York, Basic Books, 2000).

12 Arthur I. Miller, *Insights of Genius: Imagery and creativity in science and art* (New York, Copernicus, 1996).

13 Robert Garnet, 'Britpopism and the Populist Gesture', in Duncan McCorquodale, Naomi Siderfin and Julian Stallabrass, eds., *Occupational Hazard: Critical writing on recent British art* (London, Black Dog Publishing, 1998), p. 23. The book contains some rigorous criticism of the contemporary British scene.

14 Paul Bonaventura, 'A Want of Grist', in *Modern Painters*, 11 (4), Winter 1998, pp. 80–1. See also Paul Bonaventura, 'When artists lose the plot', in *Times Higher Education Supplement*, 27 November 1998, and 'Cut off from the real world', in *The Guardian*, 19 January 1999.

15 Jonathan Lear, *Open Minded: Working out the logic of the soul* (Cambridge, MA, and London, Harvard University Press, 1998).

16 Charles Jencks interviewed by John Kelleher in 'Building cosmic visions', in *Times Higher Education Supplement*, 6 August 1999, p. 20.

17 The Segsbury Project, 1998, was facilitated by The Laboratory at the Ruskin School of Drawing and Fine Art, University of Oxford.

18 See Lewis Wolpert, *The Unnatural Nature of Science* (London, Faber and Faber, 1992).

19 *Ibid.*, chapter 2, 'Technology is not Science'.

20 Richard Gregory, 'Beyond the looking glass', in *Times Higher Education Supplement*, 27 November 1998.

2 The Scientist's Mind: The Artist's Temperament

Siân Ede

'Transcendental,' said the technician
'to stumble on a quark that talks back.
I will become a mystagogue, initiate
punters into the wonder of it for cash.'
'Bollocks,' said the quark, from its aluminium
nacelle. 'I don't need no dodgy
crypto-human strategising my future.
Gonna down-size under the cocoplum
or champak, drink blue marimbas into
the sunset, and play with speaking quarklike
while I beflower the passing gravitons.'

JO SHAPCOTT, 'Quark'.[1]

Statements of fact: fragments of fiction

> 'Modern physics seems to demonstrate, beyond all reasonable doubt, that all the complexities of the cosmos will ultimately be explained in terms of a few forces and particles.'
>
> 'Some Comfort Gained from the Acceptance of the Inherent Lies in Everything.'

It would not be difficult to guess which of these statements was made by a contemporary scientist and which by an artist. The sentiments stand out as typical in every way of what we have come to regard as the confident determinism of science on the one hand with the mischievous nihilism of art on the other. The first assertion comes from Lewis Wolpert, embryologist and science writer, whom one would have to invent if he didn't already exist, as a somewhat Shavian self-ironic model of reductionist materialism. The second quotation comes from artist Damien Hirst as a title for one of the works in his notorious *Natural History* series, in which two cows are sliced into twelve vertical sections and displayed in walk-round tanks of formaldehyde. If observers wish to see in this work some sort of satire of a classic scientific process, then they are free to make that association, but this may or may not be the artist's intention.

The writer A.S. Byatt contemplates Hirst's tiger shark which swims in formaldehyde and is entitled *The Physical Impossibility of Death in the Mind of Someone Living*:

> I am very moved by it, I think because it is a case where the *objet trouvé* and the language and the framing of a thing to label it as a work of art, when otherwise it wouldn't have been, have actually *worked* and worked together – you do look at the sides, the skin, the flesh, the eye of the beast, as a beast, and as a representation of an idea – and this is owing to the mysteriously right and succinct words. It's not only the fact that it's real – it is also the fact that it's very dead – *but swimming apparently* – the idea of death swimming…[2]

This is a searching response and it may help others to see the work afresh and articulate their own reactions. But meaning is not fixed. Hirst has remarked: 'Sometimes I have nothing to say. I often want to communicate this.' If the ultimate goal of contemporary science is a single irrefutable Grand Unified

Damien Hirst, *The Physical Impossibility of Death in the Mind of Someone Living*, 1991. Tiger shark, glass, steel, 5% formaldehyde solution, 213 x 518 x 213 cm. Courtesy of the artist and the Saatchi Gallery, London. Photo: Anthony Oliver.

Theory, contemporary art anticipates not one but multiple, unfixed interpretations. These depend on the observer's personal disposition, on particular contexts, on time and place. Artworks can present meanings which are deep or slight. Hirst's pickled cow and calf might have been regarded as no more or no less than a cheeky ironic joke by diners at the London restaurant where it was part of the décor. Art is whatever you want it to be. As Jonathan Miller told Victoria Gillick on a BBC *Late Show* in 1989 devoted to the scandal of Rick Gibson's artwork earrings which he had made from freeze-dried human foetuses, 'Madam, in the late twentieth century if you call it Art it's Art.'[3]

As we shall come to see, Hirst and his fellow contemporaries *do* have something to say. And science is not as infallible as Wolpert's remark suggests. This is not to make claims for an easy compatibility, however. It does not take long in any discussion about art and science before the concept of 'creativity' is raised. Surely art and science are fundamentally the same, it is said, since in order to progress beyond routine practice both require imaginative leaps of original thinking. But Lewis Wolpert sees this as 'romantic delusion', members from each constituency suffering from a kind of genius envy. For the process of creativity in art and science is applicable to almost any human activity. 'Creativity', as a property of the ever-flexible human mind, is essential to survival and perhaps some people possess more of it and certainly all of us would benefit from operating in circumstances where it can flourish. In examining the *cultures* to which art and science belong in the contemporary world, however, it is hard to see many similarities and it is the culture clash that is so fascinating to observe when considering whether artists and scientists can learn from each other or even go so far as to collaborate. Scientists, whatever their field of study, are governed by 'the scientific method' and in investigating how the world operates theirs is a shared search for agreement; contemporary artists work alone, they make things up and encourage individual or even dissenting responses.

The 'genius' is exceptional in science but the licence enjoyed by contemporary artists to give expression to their unique individual experience suggests that any one of them might lay claim to prodigy. Indeed, fame is the spur for a great many artists leaving college, an idea exploited ironically by the artist Gavin Turk who for his degree show in 1991 at the Royal College of Art presented *Cave*, a studio empty except for a blue ceramic English Heritage plaque reading: 'Borough of Kensington/Gavin Turk Sculptor/Worked here 1989–1991'. 'This simple comment on the myth of the artist,' writes the critic Louisa Buck, 'the exclusivity of creativity, reputation, and the death of art in general was lost on the authorities of the Royal College who decided that Turk had displayed insufficient work to pass his MA degree. His response was to christen the plaque *Relic* (1991–3) and to mount it in a reverential glass case to denote its "significance" as an artwork and its symbolic role as a salute to all the artists – Joseph Beuys in particular – who have mythologised their own lives and created their own legacies.'[4] It would be hard to conceive of a scientist trying to get away with a similar act of iconoclasm. The expression of irony is simply not part of their practice. Imagination plays its part, but scientists mostly earn their reputation through extraordinary patience. A typical working scientist with a string of academic qualifications and an apprenticeship of many years may be engaged for decades in a pioneering but esoteric area of research into – say – the operation of ion channels in the beta cells of the pancreas. Working with a team of researchers he or she will know that there are other teams in other countries undertaking similar research, competition is fierce and the internet scanned daily. If the team makes a breakthrough they will be recognised in their particular branch of molecular biology and will see their names in print in learned journals ('Golgi, F.M., *et al.*'). But while there are thousands of hard-working, self-effacing artists who may never be 'discovered', most seek eventually, if not the objects of Philip Larkin's mordant envy – 'the fame and the girl and the money all at one sitting' – at least public acknowledgement that their work, embodied in their name as a mark of individuality, has made a real impact on people's emotional sensibilities.[5] The act of communication in itself is one of art's main reasons for existence.[6]

The laboratory and the art room

Most of us understand the art/science divide from our school days. The 1988 National Curriculum established science – and also maths – as one of the three compulsory subjects to be studied up to the age of 16. If this came about as part of a vague desire to educate a cohort of scientists in order to enable Britain to compete in the world's market-place, it has at least ensured that everyone, in theory, will acquire a basic scientific vocabulary. Unfortunately the arts, apart from literature, are not seen as essential and, after often superficial attention up to the age of 14, they become options only.[7]

Some aspects of current science education are good. The majority of 16 year olds should have a grasp of basic genetics, of electromagnetism and the periodic table by the time they leave school. Few science teachers are nowadays called 'Stinks'and although school science still has an old-fashioned feel to it, the school laboratory is a version of professional or academic laboratories. Potentially dangerous places where an attitude of seriousness and responsibility is required, they are domains where certain protocols are set down and a particular language is spoken. Science does develop some personal skills such as the 'curiosity, motivation, teamwork and the ability to communicate' noted in the Royal Society's 1999 statement on Science in the National Curriculum. But the main currency of that communication is logical thinking, mathematics, the application of graphs and the writing up of work, usually in numbered paragraphs, all to serve one of science's fundamental concepts, 'the role of evidence'. And although science teachers may have engaging or eccentric personalities themselves, and while living forms, including human living forms, are studied in some depth, human interests, passions, and emotions are set aside, and concerns about the impact of scientific advances on our lives are largely related to 'how society takes account of scientific evidence in making decisions'.[8] The full range of ethical considerations concerned with science – with genetics, nuclear power or sexual behaviour, for example – are more likely to be dealt with in English or, absurdly, religious education, and have only recently been suggested for inclusion in the science curriculum. So starts the general perception, articulated most vehemently by Lewis Wolpert, that 'science is value-free'. Science is about things and how they work. Amazingly, for many secondary first-year students the study of science still begins with that ineluctable object of desire – the Bunsen burner. This shows a remarkable failure of imagination, considering the large numbers of primary school children (and younger) who have a passionate interest in dinosaurs or planets or computer games. Science is serious, methodical and incremental (one learning step leads to another). Its basic method for the examination of any phenomenon is to narrow down the parameters for scrutiny and record data systematically, often just along two axes on a graph. Controlled experiment is the prime observance, although at school level it is not original experiment, because the teacher knows the results and individual opinion is inadmissible. Art rooms, in complete contrast, are places where personal expression and individual sensibilities are fostered, where, unless the teacher is unusually dogmatic, there is no 'right' way of doing anything and where a kind of glorious mess prevails. The science department and the art room are a universe apart.

Reason and feeling

Science did not emerge as a term in the modern sense until the 1830s, 'natural philosophy' being the phrase formerly used to describe a curiosity for the natural world. In retrospect, however, we can see a turning point towards modern science in the brilliant career of Sir Isaac Newton (1642–1727) who, while a serious alchemist, discovered the basic characteristics of many physical phenomena which are still fundamental to our understanding of nature. He discovered the laws of motion and gravitation and the science of mechanics, and developed the calculus. Newton made the first truly successful reflecting telescope and was the first to use a prism to split light into the colours of the spectrum. He also studied heat, acoustics and fluids.

With Newton as a figure of inspiration in England, the great age of science was the Enlightenment, the grand intellectual movement of the eighteenth century which promoted rational enquiry. This derived as much from a desire to throw light on the tyranny of superstition as it did from a curiosity about the natural world. The search for a method to investigate the natural world in a fair and reasonable manner was born out of a desire to find a 'truth' uninfluenced by any kind of prejudice or idolatry. It is interesting to note how as it emerged from centuries of darkness and ignorance, the scientific vision had a generally philanthropic purpose, where justice and tolerance formed part of a reverence for rational thought.[9] However, freedom from oppression was also the clarion call of the opponents of Enlightenment science, the Romantic philosophers, writers and artists.

Artists today are, unwittingly or not, still greatly influenced by the sentiments of Romanticism, which articulated the role of the artist as divine messenger. The idea is a very ancient one, perhaps as old as art itself, for it may be no coincidence that art probably emerged in early humans at the same time as religion, from between 60,000 to about 30,000 years ago. Both were dependent on the sense of a fantastical world hidden from the here and now, a world where past, present and future coexisted and where enemies could, at least symbolically, be overcome – the real enemies embodied in predators and tribes of foreign hunter-gatherers, and the final unknown enemy, death. Terror was expressed and danger subdued through anthropomorphism, ritual, dance, drama and depiction. Both art and religion rely on conveying mental representations of images of another metaphysical or imagined world through symbolic language and pictures.[10]

All language is symbolic, but in subjecting all received ideas and beliefs to the light of reason and empirical proof, the rationalists of the Enlightenment sought universal terms of agreement in order to discover the nature of reality. They still – in the main – acknowledged the existence of a God but the goal was to marvel at His works through a better understanding of their material properties. Perhaps the Romantics' reaction was inevitable. In the paintings, poetry and writing of the time the passions and spiritual longings of the individual were given full expression in forceful opposition to what was perceived as the cold oppression of pure intellect or reason. As critic Peter Kitson observes, it can be misleading to generalise about the range of opinion expressed over the whole period now recognised as 'Romantic', from around 1780 to 1830, but, simply put, Romanticism opposed every tenet of the Enlightenment: reason with emotion, objectivity with subjectivity, control with spontaneity, limitation with aspiration, empiricism with transcendentalism, society with the individual and order with rebellion. For Coleridge it was the Imagination which was 'the living power and prime agent of all human perception, a repetition in the finite mind of the eternal act of creation in the infinite I AM.' The Romantics particularly resented the Enlightenment demystification of nature. William Blake's opposition can be most vividly seen in his famous depiction of Newton. Sitting on a rock examining the floor on which are inscribed geometrical figures, measuring the base of the triangle with a pair of compasses, he ignores the wonder of the universe around him to concentrate on abstract reasoning.[11]

William Blake, *Newton*, 1795/*c.*1805. Colour print finished in ink and watercolour on paper, 46 x 60 cm. © 1998 Tate Gallery, London, all rights reserved.

The Romantics made many declarations which still have a ring of familiarity for artists: 'Only art can bring us close to the inexpressible', 'Man is only free when he plays, because he makes up his own rules', 'Poets are the unacknowledged legislators of mankind'. Goethe's novel *The Sorrows of Young Werther*, wildly popular in Europe in the 1770s, ends with the hero shooting himself for love in Romantic despair, in an expression of passionate resignation that is reminiscent of Kurt Cobain of Nirvana. When the suicide rate rose as a result, the book was banned in some countries and Goethe had to write a prologue in subsequent editions warning the reader not to imitate his hero. In real life reason ought to prevail.

Or should it? The contradictions embodied in the Classical/Romantic divide seem grounded in human nature and our survival depends on an application of both – reason and feeling, restraint and passion, practicality and otherworldliness. They also co-exist in the social world. But the balance was truly knocked out of kilter when, following on from Lyell's geological speculations about the vast age of the earth (a great deal older than the 6,000 years calculated by devout theologians), Darwin made public his 'dangerous idea'.[12] The intellectual climate was, in some ways, ready for a catastrophic change and some saw the new science as reinforcement for their growing doubts about the nature and existence of God. Many, however, had an underlying sense of foreboding that an anarchic bleakness would result from living in a universe without spiritual direction.[13] Matthew Arnold's poem 'Dover Beach' (1867) was not particularly written with Darwin in mind, but the hopelessness it expresses seems characteristically post-Darwinian. After harking back to a mythical time when the 'sea of faith was [once] at the full', he goes on to recognise that the only redeeming feature in the universe may be human love:

> *Ah, love, let us be true*
> *To one another! For the world, which seems*
> *To lie before us like a land of dreams,*
> *So various, so beautiful, so new,*
> *Hath really neither joy, nor love, nor light,*
> *Nor certitude, nor peace, nor help for pain;*
> *And we are here, as on a darkling plain*
> *Swept by confused alarms of struggle and flight,*
> *Where ignorant armies clash by night.*[14]

Science alone cannot be blamed for a decline in religious faith but it was – and still is – regarded as instrumental in diminishing our respect for the mystical and transcendent aspirations of the human mind, expressed so effectively by poets and painters. And it consequently affects the way we determine the principles for conducting our public and political life.

The two cultures

The debate best remembered for bringing this controversy to public attention in the twentieth century took its name from the 1959 Rede Lecture at Cambridge University. Called *The Two Cultures and the Scientific Revolution*, it was given by a man who was both a scientist and a novelist, the writer C.P. Snow. The revered English Literature scholar F.R. Leavis retaliated, though as much as three years later, with the Richmond Lecture, *Two Cultures? The Significance of C.P. Snow.*[15] The debate makes fascinating reading as a piece of social history, as much about class and power as it is about intellectual life. But it is best remembered for Snow's strident assertion that two distinct cultures existed amongst the educated elite. Writing as both scientist and novelist, he explained:

> I felt I was moving among two groups who had almost ceased to communicate at all, who in intellectual, moral and psychological climate, had so little in common that instead of going from Burlington House or South Kensington to Chelsea, one might have crossed an ocean.[16]

Snow's charge was that the non-scientific intelligentsia were out of touch with a modern industrial society, 'in a deep sense anti-intellectual, anxious to restrict both art and thought to the existential moment.'[17] They could not cope with the simplest concepts of pure science and regarded applied science as grubby. He presented a controversial challenge to what he saw as the backward-looking anti-industrial pastoralism of the establishment culture. The urge for material progress, he pointed out, brought out the best in mankind: 'common men can show astonishing fortitude in chasing jam tomorrow. Jam today, and men aren't at their most exciting: jam tomorrow and one often sees them at their noblest.' And with post-war optimism, he asserted his belief that the products of science could alleviate all the world's problems – 'scientists naturally have the future in their bones'.[18]

Coming from a tradition where the discipline of university English was 'a safe haven for professors, who had become a clergy without dogma, teaching sacred texts without God', Leavis made a stinging attack on Snow's own person and pretensions as a writer, particularly on his gung-ho linguistic style.[19] He also defended the prerogative of the literary intelligentsia over the scientific community in its domination of the British establishment.

There are sentiments in Snow's lecture that we can still recognise at a glance if we look at the two cultures today. Move from the brisk, focused, no-nonsense teamwork of the laboratory to the inconsequential allusiveness of the artist's studio, or stride the few steps along Carlton House Terrace from the 'serious' Royal Society to the 'cool' ICA, and Snow's 'gulf of mutual incomprehension' can still be perceived.[20] It would be interesting to pose again to arts people Snow's smart science questions. 'I have asked the company how many of them could describe The Second Law of Thermodynamics. The response was cold: it was also negative…'[21] More arts people may know the answer than was the case in Snow's time and knowledge about the eventual heat death of the universe may bring with it a sense of awe, but it seems far removed from life as it is lived by humans in the here and now.

For all that Leavis represents an outworn faith by divine right which held what he called 'the transmitted culture' as its holy writ, his defence of 'literature' (for which you could substitute 'art', with 'artwork' for 'poem') contains statements which still have resonance in explaining its power and purpose. Declaring contempt for Snow's expressions of empty materialism, Leavis asks fundamental questions:

> What for – what ultimately for? What, ultimately, do men live by? – the questions work and tell at what I can only call a religious depth of thought and feeling … It is in the study of literature that one comes to recognise the nature and priority of the third realm … the realm of that which is neither merely private and personal nor public in the sense that it can be brought into the laboratory and pointed to. You cannot point to the poem; it is 'there' only in the re-creative response of individual minds to the black marks on the page. But – a necessary faith – it is something in which minds can meet. [22]

Leavis does not point out, though he might have, that Snow's confidence in the benefits of industrialisation was born out of a very recent past. For science had proved itself in the winning of the war. There had been feats of engineering in war transportation, a harnessing of radio waves for broadcasting and radar, an appliance of mathematics for code-breaking and information technology, and many breakthroughs in the refinement of weaponry, culminating with a manipulation of subatomic particles for destruction on a massive scale. Any expressions of mistrust would have been unthinkably anti-patriotic at the time and would have alienated his audiences further. But he recognises the imperative for the future. 'The advance of science and technology means a human future of change so rapid and of such kinds, of tests and challenges so unprecedented, of decisions and possible non-decisions so momentous and insidious in their consequences, that mankind – that is surely clear – will need to be in full possession of its full humanity.'[23]

Although Snow tried to make a case for 'the intellectual depth of the scientific edifice, its complexity and articulation, the most beautiful and wonderful work of man', one is struck mostly by a sense that he feels it is a good thing because it will solve the material problems of the world and that is sufficient in itself, as if the political will to make this happen were not the first requisite. This leaves little place for Leavis's vision of art as a great expression of transcendent feeling and moral value. Art gladdens the offices of the powerful and is celebrated by any state in the act of reinventing itself but it does not provide guiding principles for public life. Leavis lost out not just because he spoke in patrician tones but because it had become difficult to find any words to describe the power and significance of art without sounding like a rather querulous Sunday school teacher. You simply cannot compare 'a religious depth of thought and feeling' with 'jam'. As the debate continued, however, it took a new, more complicated turn. For Snow's conviction of the benefits of scientific progress and Leavis's faith in the implicit moral value of great works of art were both to come under the guerrilla scrutiny of the philosophers and sociologists.

Things fall apart; the centre cannot hold[24]

Students of art and culture will understand the wider movement of change in twentieth-century Western thought, away from any faith in a unified fixed order towards a de-centred, godless and relativist archipelago of beliefs. Post-modernist thinking has revolutionised the intellectual world of the arts and humanities.[25] The term 'post-modern' is a loose one with no consensus of meaning, drawing as it does on a host of theories, some of them direct legacies of Romanticism, and including post-structuralism, psychoanalysis, Marxist philosophy, post-colonialism and more. In essence it questions the notion that there is a fixed and universal truth for all humankind, everywhere and at all times. In its regard for art, it denies the existence of a dispassionate reality beyond the artist's individual representation. With this in mind it examines the social, political, geographic, economic and historical context of the act of creation and also that of the reader or viewer who brings continually changing meanings to the unfixed work. The artist may work intuitively but the work itself, in theorist Roland Barthes's words, is 'always, already, written' because of its precise genesis in person, time and place. Post-modernists try to avoid making statements which imply distinctions of value or worth and in doing so accurately reflect a twentieth-century mood of moral uncertainty and a political distrust of privilege for one culture over another, seeing all cultures, including 'high' culture and 'popular' culture, as relative. In this environment 'anything goes', everything is ripe for question and it has therefore become inevitable that critical thinkers should question the nature of faith and also of so-called 'rational thinking'.

The language, or 'discourse', of post-modernism is obscure for most people but its principles are actually at large in the wider world, at least in the West, where we value freedom of choice and freedom of expression, or at any rate the illusion of them, in almost every aspect of our lives. We may also recognise that we speak many languages in the course of our private and social lives and that rational language is recognised as the language of negotiation in public. Why should this be? Post-modernist critics and cultural relativists have not been slow to 'deconstruct' the languages and practices of such discourse, especially in relation to the culture of science.

At the heart of the discourse of science is 'the scientific method'. According to the philosopher of science Karl Popper, one can never prove a theory is true because to do so one would have to make an infinite number of predictions and only a limited number could ever be tested.[26] So in order to be scientific, a hypothesis must undergo a series of empirical tests which aim to make it *falsifiable*. If it can be shown to be false, then it is rejected and new hypotheses may emerge until one is found to be unfalsifiable, and therefore, possibly, valid. Popper's system is generally acknowledged as useful by the scientific community although they recognise its limitations. Checks must be made to ensure that the experimental process itself is sound and the data interpreted accurately. Creative thinking is acknowledged but so is the fact that it may lead to pitfalls. And even where hypotheses are accepted, absolute certainty cannot be taken for granted and a precise mathematical analysis of likelihood and risk is undertaken before the scientist moves on to the next step. In practice, then, scientists operate in a culture not of explicit certainties, but of doubt and question, less with the aim of securing absolute truth and more with pursuing the theories which best explain the phenomena under examination. These should have a broad scope, in order to take into account a range of behaviours, but should also be as simple as possible, with a minimum number of hypotheses. Peer-group assessment is critical and results are published only after exhaustive observation, using blind controls and a precise capturing of data to arrive at a result which appears to be accurate 'beyond all reasonable doubt'. The analogy with criminal investigation is a useful one. However, even if a conviction is secured, based on strong corroborative evidence and after eliminating many false hypotheses from the enquiry, there may be no ultimate means of certainty and it is possible that the judgement may be adjusted or even overturned at some future date.

It is this notion which is taken up by the philosopher of science Thomas Kuhn, whose book *The Structure of Scientific Revolutions* has had an influential effect upon post-modernist observers.[27] Kuhn takes the view that most science is routine or 'normal science' which adds more evidence for a theory already part-established, a resolution of finer points, an adjustment of detail. Most new theories develop from existing ones and fit neatly into a network of existing inter-reliant theories, forming an established *paradigm*. Occasionally, however, and dramatically, particularly when observations show unexpected results, a scientist might have a brilliant insight. New inconsistent evidence may stimulate the powers both of reasoning and of the imagination. And suddenly a new theory is formulated, and sometimes more than a theory, a whole new set of propositions, which, when tested by the available data, overturn the former set of assumptions, causing a *paradigm shift*. Major paradigm shifts can have startling implications, not just within a particular area of science but also in the wider world, where they may set in motion a complete revision of ideas about the way we regard our existence. When Galileo built the first effective telescope he was able to provide real evidence for Copernicus's speculation that the earth moved around the sun. The consequences of this for science, but more spectacularly for Western thought, were enormous. If the earth was no longer at the centre of the universe, what did that say about the traditionally accepted pre-eminence of man? When Darwin published *The Origin of Species* he provoked anguished opposition from the Church because if the human race was not privileged through Creation, the universe could be seen to operate without God. Our world view was forever

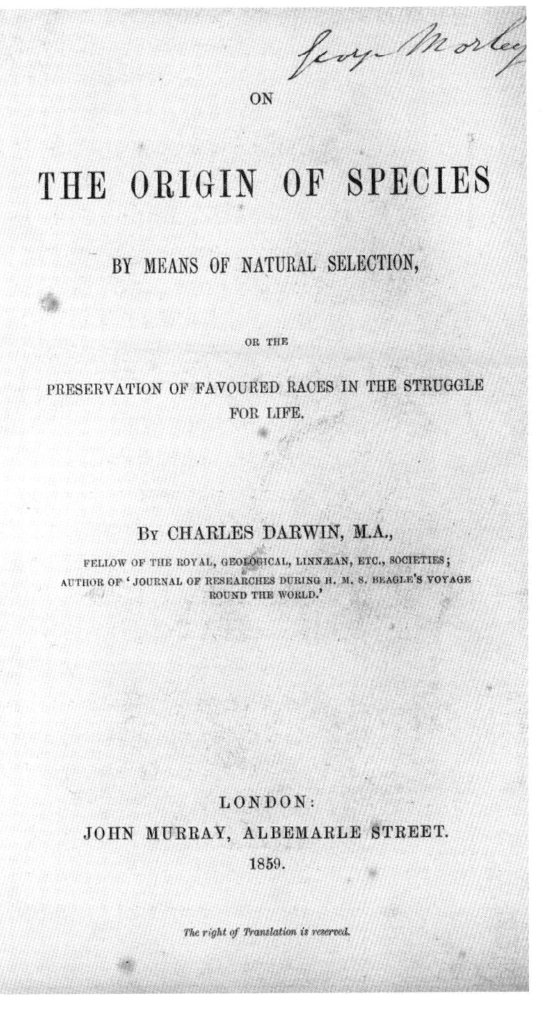

ON

THE ORIGIN OF SPECIES

BY MEANS OF NATURAL SELECTION,

OR THE

PRESERVATION OF FAVOURED RACES IN THE STRUGGLE FOR LIFE.

BY CHARLES DARWIN, M.A.,

FELLOW OF THE ROYAL, GEOLOGICAL, LINNÆAN, ETC., SOCIETIES;
AUTHOR OF 'JOURNAL OF RESEARCHES DURING H. M. S. BEAGLE'S VOYAGE ROUND THE WORLD.'

LONDON:
JOHN MURRAY, ALBEMARLE STREET.
1859.

The right of Translation is reserved.

Title page of *The Origin of Species by Means of Natural Selection* by Charles Darwin. London, John Murray, 1859. Courtesy of the Wellcome Library, London.

changed and we are still taking stock of the profound significance of evolutionary theory.

While today's scientists recognise the principle of paradigm shifts and are even prepared for surprising results in their daily practice, they cannot work without an expectation that physical laws and forces will operate as normal, and that established theories will continue to hold true. And some apparently conflicting paradigms, even major ones, can co-exist and be applied, if not simultaneously, certainly at different scales – macro and micro – for different purposes. So Newton's laws are still applicable in the macro-world and although there is as yet no simple reconciliation between Einstein's General Theory of Relativity and current views on the indeterminate behaviour of subatomic particles in the quantum world, physicists can operate at both levels. And a shift in perspective, arrived at through examining new evidence, does not mean that there is also a shift in fundamental reality.

However, for some contemporary thinkers, paradigm shift theory has set in train radical doubts about the possibility of any grasp on certainty. And the very methodology itself has also come into question. How far can anyone rely on premises which might turn out to be false, or data analyses based on shaky hypotheses? If theories can never be ultimately verified and paradigms are unstable, then surely the credibility of the whole practice comes into question? The philosopher Paul Feyerabend takes this doubt to its extreme. 'The idea that science can, and should, be run according to fixed and universal rules, is both unrealistic and pernicious,' he writes. 'All methodologies have their limitations and the only "rule" that survives is "anything goes".'[28] Anything goes? This is a surprising sentiment to apply to the scientific method. For what, then, distinguishes a scientific proposition from one in any belief system? If 'anything goes', then surely we are free to compare science with mythology or religion?

It is now only a short step for critics to assert that if all paradigms, scientific or otherwise, are a matter of belief, then no one belief system can take precedence over another. This will apply to individual religions which, as belief systems, have a long history of incommensurability with the rationalism of science, as they also have with each other, being impossible to compare because they do not have common frames of reference. All beliefs or belief systems, great and small, political, superstitious, alternative or based on myth or personal predilection, may be equally valid. None can claim privileged knowledge. Astrology, Feng Shui, and the Tarot have a popular following, as did the phrenology, witchcraft trials and doctrines of magical sympathies and antipathies of the past. Context is important, of course, but who is to say whether one context should be more privileged than another? For if 'anything goes' the consequence is probably, 'whatever you want' and the ascendancy of one belief system over another may simply be a matter of personal choice, as it is in the pick-and-mix culture of the relatively free Western world. Or, if not personal, it is predicated by the prevailing culture. It is interesting to note, for example, that some American states and school districts have tried to rule that the teaching of evolutionary concepts in science classes should be balanced with instruction in biblically orthodox accounts of human creation, with the state of Alabama mandating that science textbooks describe evolution as no more than a theory.[29] And creationists in Kansas have gone one step further and have sought to exclude altogether references to evolutionary theory from the science curriculum.

Scientists have naturally been outspoken about the relegation of the scientific method to a league of the meretricious. It is one thing to have the methodology of your profession questioned but quite another to have the whole enterprise, with its resolute challenge to ignorance and superstition, completely undermined. They have expressed particular concerns about the potentially dangerous consequences of belief in something not ascertainable by the methodologies of evidence and reason. Although some scientists have religious faith, they appear to be in the minority and most openly loathe anything remotely obscurantist, such as astrology, alternative medicine or folk psychology. Some are not even prepared to encounter mildly entertaining diversions featuring the supernatural world, regarding them as the top of the slippery slope to delusion and exploitation.[30] But this has not deterred the sociologists of science who have treated the laboratories as sites for anthropological discovery, noting the peculiar credos, habits and language of the inhabitants and declaring that science like any other culture is the product of social attitudes and prejudices.[31]

In 1996 an article appeared in the American cultural studies journal *Social Text* by physicist Alan Sokal entitled 'Transgressing the Boundaries: Towards a transformative hermeneutics of quantum gravity'.[32] Drawing on the evidence of indeterminacy in quantum theory, it questioned the concept of an external world with properties independent of human life. It declared that physical 'reality', like social reality, was essentially a social and linguistic construct and that the universal precepts of science, such as Euclid's *pi* and Newton's G, could not be separated from 'their ineluctable historicity'. In the manner of such discourse Sokal cited prestigious intellectuals such as Derrida, Lacan, Irigaray and Lyotard, quoting from works where they had appropriated scientific theories and given them alternative interpretations.

Not long after the article appeared in print its author spoke out. It had been a hoax, produced with the aim of illustrating to the world in general that in the case of post-modern theory and social constructivism, particularly in respect of its subversion of science, the Emperor wasn't wearing any clothes. The affair provoked a clamorous international response. The editors of *Social Text* were derided for publishing a paper without subjecting it to sufficient scrutiny, so that its many errors of logic and fact, some of them obvious even to high school science students, were accepted as read, because, as the author explained, the editors were ignorant of science yet clearly believed that scientific language conveyed an impression of intellectual *gravitas*. The article had been accepted without question because it supported the prevailing political and cultural orthodoxy of the journal. In one stroke Sokal expressed for many scientists – and indeed for some in the arts and humanities – a profound weariness with post-modernist theory in general. Comment and correspondence ricocheted throughout the academic world. Alan Sokal and Belgian physicist Jean Bricmont went on to publish a book, *Intellectual Impostures* (first in France, later in the English-speaking world).[33] This contained chapters devoted to ridiculing the liberal misuse of scientific concepts and language in the theoretical discourse of many of the most revered post-modernist thinkers and writers. Backed up by a thorough analysis of the worst excesses of contemporary 'epistemic relativism', it spoke for the science community in its revulsion for the obfuscation, self-reference and self-reverence of much post-modernist discourse.[34] The response of the (mainly French) intellectuals who had been the main target of the spoof was summed up simply in the words of Jacques Derrida: 'Le pauvre Sokal'.

In a debate with Sokal at the London School of Economics in the spring of 1998, sociologist of science Bruno Latour walked into a lion's den of a lecture theatre which appeared to be packed to the corridors with triumphant scientists and a few defensive humanities academics. It soon became apparent that Sokal and Latour inhabited different universes. They appeared unable to apply the same meanings to the same words and could not agree on distinctions between definitions of the 'world out there' and our knowledge of it. In some ways this was a new debate because it was applied to science and post-modernism; in others a rather old one, being another illustration of the deep chasm between the (mainly) Anglo-American analytic tradition in philosophy and the 'Continental' linguistic one.[35] And if Sokal acted as if he had downright common sense on his side, Latour had sophistication, wit and complexity on the other.[36]

The debate is essentially about what we mean by 'the real world'. Post-modernist thinkers believe our view of the 'real world', indeed, 'real *worlds*', is a socially and linguistically constructed version. In the examination of natural phenomena, it depends where you stand: nothing can be completely isolated from a human context. Science practitioners, on the other hand, operate on the basis that there is a coherent natural world, 'in reality', irrespective of human culture. Is it possible to examine how far human considerations affect scientific research?

Versions of the truth – cultural values in the laboratory

James Watson's book *The Double Helix*, about the race with Francis Crick to discover the structure of the DNA molecule in 1953, reads like a thriller, even to the non-scientist, who is caught up in the excitement of scientific discovery.[37]

But as a record of an event that actually happened it gives only one view of the situation. Published in 1968, by a man who was by then a great British hero, it appeared to speak with authority, and its sexual politics, though questionable now, were not unusual for the time. The way Watson describes the chaps hankering after 'popsies' as a diversion on those chilly Cambridge nights is no more nor less offensive than a Raymond Chandler novel. But his views on the character of scientist Rosalind Franklin appear surprisingly frank in a book about scientific discovery. Franklin was working at King's College London, where she was undertaking her own research into the structure of DNA along with scientist Maurice Wilkins. Crick and Watson, who were in contact with the scientists at King's, were working on speculative models. Franklin, like Wilkins, was using X-ray crystallography at which she was 'a superb experimentalist, patient, dextrous and untiring'.[38] A needle-thin beam of low-intensity X-rays

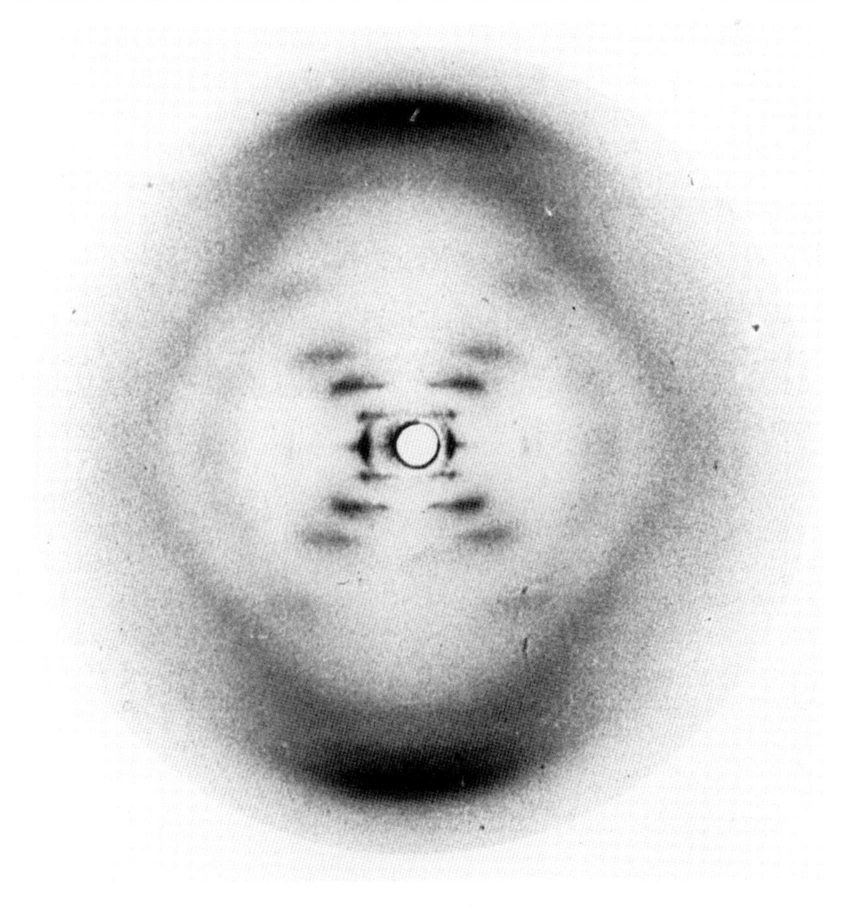

Rosalind Franklin, X-ray diffraction pattern of the B form of DNA, 1951–4. Courtesy of King's College London.

was shone though a fine fibre of DNA. Each picture took hours of exposure and the fibre had to be kept at a constant high humidity because the molecular structure altered as the fibre dried. Franklin's greatest achievement was her ability to manipulate the experiment in order to expose a pattern which was bold and simple enough to define. She was pursuing her own ideas about the likely structure of DNA. Crick and Watson were developing the double-helix model and having had sight of Franklin's data and X-ray pictures they were able to confirm their speculations. It is still not entirely clear how far Franklin was from making the connection herself.

While Watson clearly held Franklin in respect as a scientist, what stands out today are his comments on her appearance and personality: 'by choice she did not emphasise her feminine qualities ... she was not unattractive and might have been quite stunning had she taken even a mild interest in clothes. There was never any lipstick ...'[39] He 'could not see any decent way to give her the boot ... there was no denying she had a good brain. If only she could keep her emotions under control.'[40] Of course, emotions might also be expressed by a man, but Watson implied hers were typical of a certain type of serious woman and his conclusion was that 'the best home for a feminist was in another person's lab'.[41]

Rosalind Franklin tragically died in 1958 at the age of 37 before it was decided to award the Nobel Prize to Crick and Watson, together with her colleague Maurice Wilkins. In an epilogue to *The Double Helix*, Watson diluted the force of his former antipathy by praising her scientific achievements and regretting that he and his colleagues had realised 'years too late, the struggles that the intelligent woman faces to be accepted by a scientific world which often regards women as mere diversions from serious thinking'.[42]

An inevitable consequence of Watson's book is that Rosalind Franklin has become for some feminists an icon of resignation to a woman's place in a man's world.[43] But this is a distortion too. 'Women worked as scientists at every level from top to bottom and held between a third and a quarter of the professional posts,' comments Horace Freeland Judson.[44] Dame Honor Fell, senior biological adviser to the unit at King's at the time of Franklin's stay there told Judson, 'I never saw a sign of sex discrimination. Of course Franklin was a rather difficult character ... but I don't think she was discriminated against because she was a woman.' And Judson goes on to say, 'Franklin's colleagues at King's who were women unanimously reject as unhistorical and anachronistic the use of Rosalind Franklin as an emblem for the condition of women in science.'

It is too simplistic, then, to take the sexist line. But what is clear to the outsider is the fact that laboratories, like any other work arena, are not the clinical and passionless places they might aspire to be, but are settings for the whole range of emotions that are in evidence when human beings interact, and that this can and sometimes does affect the progress of the work in hand. If

Watson and his colleagues had invited Rosalind Franklin to join their team – and she had been prepared to join – it is possible that they would have arrived at their conclusions sooner. But can these human interactions also affect the nature of the phenomena under scrutiny or shake a belief in the 'real world'?

Some critics claim that they can and that human predilections can determine or alter scientific concepts and findings in themselves. Some areas of study may even have a gender, and the generally male ethos perceived to prevail in laboratories means that some are privileged over others. The critics Luce Irigaray and Katherine Hayles, for example, believe that women, by their very nature and biology, understand fluid mechanics better than men, but that we have made more progress in solid mechanics because men have chosen to prioritise a study with which they feel a closer psychological empathy. 'In the same way that women are erased within masculinist theories and language,' writes Hayles, 'existing only as not-men, so fluids have been erased from science, existing only as not-solids. From this perspective it is no wonder that science has not been able to arrive at a successful model for turbulence.'[45] Irigaray even claims that Einstein's famous equation $E=mc^2$ is a sexed equation, 'privileging that which goes fastest'.[46]

This is an interesting thought although it does sound not unlike the perception, allegedly held in Rosalind Franklin's time, that women had a natural talent for X-ray crystallography because they possessed an intuitive ability, less developed in men, to see three-dimensional chemical shapes in the patterns, besides a 'natural' manual dexterity and a willingness to undertake tedious mathematical calculations.[47] It depends on your viewpoint whether you see this as sexist, as a celebration of women's particular talents which did assist with a breakthrough in science, or, again, as totally irrelevant.

And yet it moves[48]

Needless to say, such deliberations are treated with contempt by most in the science world. Some dissenting voices will point out that our social attitudes determine how we *present* the findings of science, so, for example, the invented notion of the 'selfish' gene might be used to explain and justify certain imperialist tendencies in human behaviour.[49] This is certainly something of which we should be wary when it comes to making social use of scientific knowledge. But it is not the same as saying that the essential findings of science are also subject to influence. The nature of 'reality', the great majority of scientists believe, is constant, and it remains the same irrespective of the personality, sex or cultural mores of the people who make the discovery. No matter what went on among the inhabitants of the King's and Cambridge laboratories in 1953, the essential structure of the DNA molecule was unaffected – and still is – by human considerations. Of course, we might one day, with better instruments of observation and more contextual knowledge, see the DNA molecule differently.

But it is not a linguistic or cultural construct. It is a discrete thing in itself, obeying the impersonal and ultimately fixed laws of nature.

Such a conviction in the immanence of reality results in scientists' sense of an 'implicate order', to quote the beguiling words of the physicist David Bohm.[50] This does not necessarily assume certainty or stability. The behaviour of subatomic particles in quantum physics, which has introduced unpredictability into a formerly mechanistic picture, is still puzzling, even though nowadays experiment agrees with theory to within a few parts per hundred million. Chaos theory tries to predict how an event of infinitesimal smallness can set in train a series of consequences of escalating disturbance which might result in catastrophic events, an idea charmingly evoked in the idea of 'the butterfly effect' where the mild flutter of the wings of a single butterfly in China might result in a hurricane on the other side of the world. Mathematical predictions can be made such that even chaos appears to have some sort of 'implicate order'. Evolution is indeterminate but it has its rules too, for its chance mutations can be seen to have been selected in response to the imperative to survive in an environment governed by ultimate physical laws and restraints. This vision, of an implicit order, coherence and constancy in the physical universe, gives many scientists a feeling of deep satisfaction and awe, as if, paradoxically, they may be able to arrive at a kind of spiritual contentment and calm through the ultimate in material knowledge. Many therefore regard their profession as an honourable calling, giving them the status of explorers at the edge of discovery on a majestic scale. 'The way I see it', writes Frances Ashcroft, Professor of Physiology at the University of Oxford, 'I am piecing

Cornelia Parker, *Einstein's Abstracts*, 1999. Photomicrograph (x 50) of the blackboard covered with Einstein's equations from his lecture on the Theory of Relativity, Oxford, 1931. Cibachrome on aluminium, 58.5 x 75.5 cm. (With thanks to the Museum of the History of Science, Oxford.) Courtesy of the artist and Frith Street Gallery, London.

together the puzzle – my aim is to see "the interconnectedness of it all" – how all the bits fit together to produce something gloriously new.'

As a result of this overall vision many scientists are clear about a function for art. It is there, is it not, to show us the intrinsic beauty of the world? Moved by wonder for the phenomena they see or think about in their daily work – the microscopic cells, the rare plants, the theoretical behaviour of subatomic particles, the elegance of mathematics – they are surprised to learn that 'beauty' is a word used sparingly by artists who want to do more than simply record or illustrate objects and ideas. Rooted in the directionless world of the here and now, artists see mostly fracture, fragmentation and disarray. They may express compassion and humanity, their work may display a poignant desire for inner meaning but each individual expression is different, as is the individual response of their viewers. There is implicit disorder. Post-modernist discourse, for all its contortions, tries to articulate this in theory: artists make it in practice.

Notes and references

1 Jo Shapcott, 'Quark', from *My Life Asleep* (Oxford, Oxford University Press, 1998).

2 A.S. Byatt, in correspondence with the author, March 1999.

3 The earrings were shown at the Young Unknowns Gallery, Waterloo. Tried at the Old Bailey, Gibson was found guilty of indecency. Fining him £500, Judge Brian Smedley told the jury, 'We are not here to set ourselves up as arbiters of taste. You are here to set the standards of public decency.' Jonathan Miller is well-known as a theatre and opera director and also as a former medical scientist, and is well-regarded for his views on both cultures. Victoria Gillick is an outspoken defender of family values.

4 Louisa Buck, *Moving Targets: A user's guide to British art now* (London, Tate Gallery Publishing, 1997).

5 Philip Larkin, 'Toads', from *The Less Deceived* (London, The Marvell Press, 1955).

6 In his investigation into the earliest forms of art, archaeologist Steven Mithen defines art as stored information using symbols which are arbitrary to their referents created with the intention of communication. Steven Mithen, 'The big bang of human culture: the origins of art and religion', in *The Prehistory of the Mind* (London, Thames and Hudson, 1996).

7 In *All our Futures*, the report of the National Advisory Committee on Creative and Cultural Education, 1999, the arts are located within a broader strategy to develop young people's creative and cultural education.

8 *Science and the Revision of the National Curriculum*, Royal Society statement, January 1999.

9 A respect for reason is not just a preoccupation of Western thought. In the Sanskrit scripture *The Bhagavad Gita*, written *c.*500 BC, 'the first condition for a man to be worthy of God is that his reason should be pure. Reason is the faculty given to man to distinguish true emotion from false emotionalism, faith from fanaticism, imagination from fancy, a true vision from a visionary illusion.' Juan Mascaro, foreword to *The Bhagavad Gita* (London, Penguin Books, 1962).

10 Steven Mithen, *The Prehistory of the Mind* (London, Thames and Hudson, 1996).

11 Kitson explains that the binary divides are, of course, an oversimplification. For this section, including the reference to Blake, I am indebted to Peter J. Kitson's essay 'Beyond The Enlightenment: The Philosophical, Scientific and Religious Inheritance', in *Companion to Romanticism*, ed. Duncan Wu (Oxford, Blackwell, 1998).

12 See D.C. Dennett, *Darwin's Dangerous Idea* (London, Allen Lane, The Penguin Press, 1995; New York, Simon and Schuster, 1995).

13 For an analysis of the decline in faith and the effects of science on it in the nineteenth century, see A.N. Wilson, *God's Funeral* (London, John Murray, 1999).

14 Richard Dawkins remarks on a similar response to his books, especially *The Selfish Gene*, 2nd edn (Oxford, Oxford University Press, 1989), with one teacher advising a tearful pupil not to show it to any of her friends for fear of contaminating them with nihilistic pessimism. His book *Unweaving the Rainbow* presents arguments to show how science should inspire rather than undermine the poetic imagination. Richard Dawkins, *Unweaving the Rainbow* (London, Allen Lane, The Penguin Press, 1998), p. ix.

15 For the original text and a recent appraisal see

C.P. Snow, with an introduction by Stefan Collini, *The Two Cultures* (Cambridge, Cambridge University Press, 1993), and F.R. Leavis, with an essay by Michael Yudkin, *Two Cultures? The Significance of C.P. Snow* (London, Chatto and Windus, 1962). For a philosophical appraisal see Tom Sorell, *Scientism* (London and New York, Routledge, 1994).

16 Snow (1993), p. 2, see note 15.

17 *Ibid.*, p. 5.

18 *Ibid.*, p. 5.

19 Adam Bresnick, in *Times Literary Supplement*, 11 December 1998, p. 11, review of *The Rise and Fall of English* by Robert Scholes (New Haven and London, Yale University Press, 1998). 'When God died in the West, the light of transcendence was not so much extinguished as redirected towards the mortal artificers who built the great works of the Western canon. So it was that the first literature professors functioned as quasi-priestly expositors who sought to inculcate in their students the requisite awe before the divine genius of Dante and Shakespeare.'

20 Snow (1993), p. 4, see note 15.

21 *Ibid.*, pp. 14–15.

22 Leavis (1962), pp. 23 and 28, see note 15.

23 *Ibid.*, p. 26.

24 W.B Yeats, 'The Second Coming', from *Michael Robartes and the Dancer* (London, Macmillan, 1921).

25 For useful guides with original texts, see David Lodge, *Modern Criticism and Theory* (London and New York, Longman, 1988), and Raman Selden, *The Theory of Criticism* (London, Longman, 1988).

26 Karl R. Popper, *The Logic of Scientific Discovery*, trans. by the author with Julius Freed and Lan Freed (London, Hutchinson, 1959). For a detailed analysis see Lewis Wolpert, *The Unnatural Nature of Science* (London, Faber and Faber, 1993).

27 Thomas Kuhn, *The Structure of Scientific Revolutions*, 2nd edn (Chicago, University of Chicago Press, 1970).

28 Paul Feyerabend, *Against Method* (London, New Left Books, 1975), pp. 295–6.

29 Edwin J. Larsen, 'In God, not scientists, they trust', in *Times Higher Education Supplement*, 11 December 1998, p. 22.

30 Richard Dawkins, in particular, is deeply concerned about the sinister effects of superstitious beliefs or any dabbling in the supernatural. See Dawkins (1998), p. 28, see note 14: '*The X-Files* systematically purveys an anti-rational view of the world which, by virtue of its recurrent persistence, is insidious.'

31 See Bruno Latour and Steve Woolgar, *Laboratory Life: The construction of scientific facts* (Princeton, Princeton University Press, 1979), and Bruno Latour, *Science in Action: How to follow scientists and engineers through society* (Milton Keynes, Open University Press, 1987).

32 Alan D. Sokal, in *Social Text*, 46/47, Spring/Summer 1996, pp. 217–52.

33 Alan D.Sokal and Jean Bricmont, *Intellectual Impostures* (London, Profile Books, 1998).This book contains an excellent survey of the epistemology of contemporary science and post-modern responses to it, some of which informs this section of the chapter.

34 E.g. '*The Postmodern Brain* by Gordon C. Globus (University of California, Irvine). This interdisciplinary work discloses an unexpected coherence between recent concepts in brain science and postmodern thought ... The non-linear dynamical brain as developed shows quantum nonlocality, undergoes chaotic regimes and does not compute. Heidegger and Derrida are "appropriated" as dynamical theorists who are concerned respectively with the movements of time and being *(Ereignis)* and text *(Différence)*. The chasm between postmodern thought and the thoroughly metaphysical theory that the brain computes is breached, once the nonlinear dynamical framework is adopted. The book is written in a postmodern style, making playful, opportunistic use of marginalia and dreams, and presenting a nonserial surface of broken complexity. Publisher's catalogue.' (Excerpt from 'Pseuds Corner', *Private Eye*, 16 October 1998.)

35 A great deal of twentieth-century epistemology has focused on making distinctions between the act of perceiving and the object perceived. The phenomenalists have contended that one cannot distinguish between the objects of knowledge and objects as one perceives them; the neo-realists that objects exist independent of one's mental state. Distinctions have developed around the schools of logical positivism – there is one kind of knowledge, scientific knowledge, and this must be verifiable through experience – and linguistic philosophy – an examination of how basic epistemological words such as 'knowledge' and 'perception' are used. For a detailed study in relation to science, see Sorell (1994), see note 15. For the positing of a 'successor language to science and politics', see Bruno Latour, *Pandora's Hope* (Cambridge, MA, Harvard University Press, 1999).

36 Naturally there are many opinions about the use of a term like 'common sense'. Science is supposed to be beyond it, see Wolpert (1993), see note 26; Marxist critics like Antonio Gramsci (1891–1937) believe that the term is used as an emblem of bourgeois values.

37 James D. Watson, *The Double Helix*, ed. Gunther S. Stent (London, Weidenfeld and Nicolson, 1981).

38 Crystals are bombarded with X-rays in order to form pictures which show the positions of the constituent elements.

39 Watson (1981), p 14, see note 37.

40 *Ibid.*, p. 15.
41 *Ibid.*
42 *Ibid.*, pp. 132–3.
43 See Elizabeth Janeway, *Man's World, Woman's Place* (New York, William Morrow, 1971).
44 For a detailed discussion see Horace Freeland Judson, 'Annals of Science: The legend of Rosalind Franklin – Was the only woman in the race to discover DNA cheated out of a Nobel prize?', *Science Digest*, January 1986.
45 Katherine N. Hayles, 'Gender encoding in fluid mechanics: Masculine channels and feminine flows', in *Differences: A journal of feminist cultural studies*, 4 (2), pp. 16–44.
46 See Luce Irigaray, 'Le sujet de la science est-il sexué?'/'Is the subject of science sexed?', trans. Carol Mastrangelo Bové, in *Hypatia*, 2 (3), 1987, pp. 65–87.
47 X-ray crystallography was a field undertaken successfully by other prominent women scientists such as Dorothy Hodgkin and Polly Porter at Oxford, and Kathleen Longside at the Royal Institution. See Janet Browne, 'Crystal Clarity', in *Times Literary Supplement*, 15 January 1999, a review of *Dorothy Hodgkin, A Life* by Georgina Ferry (Cambridge, Granta Books, 1998).
48 When Galileo publicly confirmed Copernicus's theory that the earth moves round the sun, the Inquisition forced him to recant, under threat of torture. But after his public recantation he is said to have muttered, '*Eppur si muove*' ('And yet it moves').
49 For fuller discussions of this view, see Mae-Wan Ho, *Genetic Engineering: Dream or nightmare?* (Bath, Gateway Books, 1998), esp. chapter 4, 'The Origins of Genetic Determinism'.
50 See David Bohm, *On Creativity*, ed. Lee Nichol (London and New York, Routledge, 1998).

3 Many Complex Collisions: New Directions for Art in Science

Siân Ede

... I'll never make a deluge drawing
or be gripped by the science of circular
motion. And I probably won't learn to care how
many complex collisions happen in a pool
when water is trickled from above.
How currents percuss against each other,
and how waves rebound into the air, falling
again to splash up more water in smaller
and smaller versions of the same.

JO SHAPCOTT, from 'Leonardo and the Vortex'.[1]

The various interpretations allowable in post-modernist theory have become the norm in many areas of the arts, where they have a valued function in exploring and reinterpreting, through a language closer to poetry than to logic, what is rationally inexplicable. But the discourse is exchanged principally within the citadels of the academy and it is not the language of the streets where many young artists feel more at home. Their practice may be the living embodiment of post-modern slipperiness but they don't care. That they are able to choose is a manifestation of freedom in itself. If science is a search for coherence, contemporary art has become an expression of dissent.

In a free society artists can say and do what they like. They do it anywhere and through any medium – using traditional materials such as paint, photography, stone, metal or found materials, or by electronic means, installation or performance. Their influences may be conscious and chosen, or accidental and intuitive. The 'real world' for some may indeed be based on a coherent spiritual vision, arrived at through orthodox religions or through identifying a universal metaphysic through an eclectic blend of spiritual borrowings. Alternatively, they may be manifestly political or psychoanalytical, or some or none of these things. Their work collectively demonstrates a questioning of any authoritative view, a respect for all 'cultures' (of feminist, racial and queer politics, as well as actual anthropological cultures) and an acknowledgement of the viewer in the interpretation of a work. If they have any gods in their personal credos they come from what, ironically, Leavis might still have called 'the transmitted culture', but it is not his Great Tradition approved by the keepers of a national canon. Artists look back to other artists and view their own work as a continuation of a noble project in which they feel honoured to participate. Numerous influences from world art play their part, with Marcel Duchamp and Joseph Beuys still the high priests of resistance for many.

In contemporary work artists make unexpected associations and juxtapositions, conceptually and visually. They blend, mix, blur and contrast ideas, materials and relics, taking pleasure in the paradox and the unpredictable, highlighting the ordinary, the odd and the obvious. If visual images in the public and commercial domain are polished and high-tech, artists in contradiction often make work which is ugly, incidental, low-tech and messy. Irony, parody, playfulness, wit, defiance, vulgarity, a subversion of facts and a magpie appropriation of the detritus from any cultural environment are all permissible to jolt the attention and stir the emotions, whether pleasure or amusement, fear or melancholy or rage. You can ask artists what they are doing and they will tell you; some write eloquently. But preoccupied with getting their work actually realised, they would prefer the art to speak for itself.

There are thousands of examples. Here is just one. The reader needs to experience it at first hand but in bald description it is this.

Dorothy Cross, *Chiasm*, 1998. Video still. Film of Poll na bPéist (worm hole), Aran Islands, Co. Galway, projected on to hand-ball alleys at St Enda's College, Galway. Installation involving singers Carol Smith and Eugene Ginty of Opera Theatre Company, Ireland. Courtesy of the artist and Project Arts Centre, Dublin.

Chiasm

On the Aran Islands off Galway on the west coast of Ireland there is a strange geological fault. Called Poll na bPéist (worm hole), it is situated at the base of a limestone cliff and is almost completely rectangular, with the dimensions of a small swimming pool. The artist Dorothy Cross has filmed the fault at high tide when it is filled with water.

Hand-ball alleys were built for a kind of squash game popular in Irish villages and towns for a brief period during the 1930s. They have since fallen into disuse and are now regarded mostly as weedy eyesores, places where the local youth hang out. In a bizarre symmetry, the proportions of the fault match those of the hand-ball alleys. Dorothy Cross's digitally manipulated film of the fault has been projected onto the site of hand-ball alleys at Salthill, Galway. In the darkness of late evening, two mirroring film images of the water-filled Poll na bPéist are projected from 90-foot towers to fill the floors of the alleys, creating contained oceans, constantly moving within the space. The artwork is called *Chiasm* and is an exploration of the myth of Ariadne and Theseus. (Having promised to marry her, Theseus abandoned Ariadne on the island of Dia. Left alone, she filled the air between Dia and her home of Crete with her tearful laments). Opera Theatre Company singers, tenor Eugene Ginty (whose father was a hand-ball champion) and soprano Carol Smith perform in separate alleys, walking within the projected ocean. They sing Verdi, Gluck, Berlioz, Gounod, Strauss, Puccini, Purcell, Bizet, Tchaikovsky. The project, managed by Project Arts Centre, Dublin, continues off-site via website, video and film.

How do we explain and interpret this work? Here are some fragmentary hypothetical responses:

> A configuration of cultural coincidences … Irish culture is sanctified and enjoined to the high European … rural idylls … tectonic plates … *memory and desire* … the gradual revelation of archetypes which demonstrate not just a localised but a collective unconscious … the rootedness of the ordinary … woman and nature … time-tested symbiosis of matter, energy and information processes … reconfiguring the bounds of diurnal life in tendentious connection with the past … the foregrounding of the perversely arcane and time-based before emblems which appear to have a universal prerogative, serves as an ironical observation … *male human sacrifice was integral to the worship of Ariadne … torn to pieces by a flock of delirious women intoxicated by ivy* … the suspension of autocratic, reductionist thought patterns has given rise to an arbitrary consciousness and ultimately ideational plenitudes … the urge to create a symbolic logic that draws out some kind of order and meaning from a dislocated surround … *Remember me! Remember me!…* the felt anxieties promoted through the discourse of the sea distances us

> from the sublime aspirations perceived, as Lacan observes, through dreams of an alien Other ... performative installation programmed to use technology to create a cyber-transcendent meta-world ... high camp rendered bleak in the terse wind and rain ... *Caught in that sensual music all neglect/Monuments of unageing intellect ...*[2]

There could be endless interpretations. Such commentaries may be original personal expressions, or manifestations, to a greater or lesser degree, of fashionable critical perspectives. They demonstrate what is at the heart of practice in the contemporary arts and humanities: a continual reconfiguring of ideas, a ceaseless reinterpretation of meanings.

Art in a material world

How should we talk about art? It is surprising to discover how much *writing* surrounds the visual arts, as if, in a culture which places a high value on verbal expression, the visual arts constituency needs to prove its intellectual credentials. Art students are required to write dissertations about their work and this may amount to as much as a third of their output. Not all of them can write well. Why should they?[3] Many visual arts organisations produce reams of dense text, especially since desktop publishing has become available to all. Artists' books abound. Some are new works of art in their own right but, alas, a great deal of writing about art is not very accomplished or very interesting. In style it can veer from the most esoteric and self-referential of post-modernist discourse to worthy Sunday school assertions which do not stand up to much scrutiny ('the work forces us to confront our own mortality' etc.) or, as favoured in some contemporary art catalogue essays, presents someone's personal 'poetic' response to the art as if the main purpose of that art was to be a visual stimulus solely to this end. This simply does not happen in any other art form. While there is a place for academic commentary and good programme notes, no one expects novelists, poets, dancers, musicians and theatre people to accompany their work with wordy reveries. The work speaks for itself directly. Perhaps fortunately, much art criticism is not widely read or understood. And although accessible, intelligent explanations and appraisals do exist in some contemporary catalogues and in some press reviews, even these convey a sense that they emanate from a closed and privileged world inhabited by those in the know. An air of defensiveness pervades.

Value judgements within the art community *are* made in private but it is perhaps thought that if they were to be uttered in public they might call the whole enterprise into question. There is a material reason for this which applies to the visual arts more than to any other art form, namely that many artworks are commodified. They are bought by people with money to spare – to please, decorate, enhance self-image and to serve as investments. Artists' names are 'brokerable', and critical valuations and catalogue essays can be read as share prospectuses. The international art world is rich, precious and often exclusive and, whether we like it or not, it has an important influence on the way the rest of us regard the work. When a collector can sweep into a degree show and speculate on talent, notions of price and value are conflated. And it is this bottom-line financial valuation of the work which so enrages the bourgeoisie, an effect which, of course, only serves to reinforce the value of the speculation.

Artists are ambivalent about this state of affairs. To be invited into the enclosure is flattering and, of course, financially rewarding in a career which is rarely lucrative. An artist's reputation may be made but this may pose a threat to independence and integrity. Once accepted, there is a danger that one will be condemned to keep churning out work in an easily recognisable style. Consequently, as if to bite the hand which feeds them, many artists perversely push their work to the limits: 'I wanted to find out where the boundaries were,' Damien Hirst has said, 'so far I've found there aren't any. I wanted to be stopped and no one will stop me.' In a free society where 'anything goes' this is indeed some challenge. How far can you go, and does anyone care?[4] It sounds like the cry of a child of liberal parents, wanting to test the limits of their forbearance. And as good liberal parents know (and are despised for knowing), the riot of youth for its own sake deserves celebration and, manifested in art – as in fashion, or pop music – it certainly adds to the gaiety of nations and increases the public stock of harmless pleasure.[5] And if the curmudgeons are peevish about sex, drugs and rock and roll and all the attendant fun, noise and irresponsibility, then it may be because they really *are* being reminded of their own mortality. Behind the show of bravura, however, there beats a desire to make a difference, to be taken seriously and to communicate a sense of solidarity with the rest of the human race. The making of art is not 'self-expression' for its own sake. It is too difficult and risky an enterprise for something so self-indulgent.

It is surely time for the art community to speak up for itself, to explain, as popular science writers certainly have, what their enterprise is about. Scientists are required to substantiate their claims as part of their methodology and in public, therefore, they are eloquent in making their case, even when there is controversy. The rational language they rely on is better understood by the decision-makers and there is anyway an implicit belief that what the scientist does is intellectually rigorous and has practical applications. Even potentially adverse consequences are presented in the form of plausible risk analyses. The work of artists, in comparison, is often dismissed as wilfully subjective or idiosyncratic and at the end of the day a question of personal taste. If at worst it is a fashion item, at best it is the expression of individual 'genius', something untouchable and ultimately irrelevant to public life. Audiences attend exhibitions in search of work which may move them, sources of contemplation and meaning, but it is difficult to give explanations about the power of these experiences when they do occur. It is easier to recognise the value of a blockbuster exhibition of art from the past, of Turner, say, or Monet or Picasso and now even Jackson Pollock, because their iconography has become reassuringly familiar. We need more help, more expressions of discrimination, more open vigorous debate about new art, and it is unsatisfactory when some in the art community appear to hide behind the illusion that their work is too avant garde for most of us to understand. Some of it *is* banal and trivial. But some is rich and moving and interesting and we want our eyes opened to it. The science world has recognised that it can be misunderstood by the public and has highlighted the need for positive initiatives to enhance the public understanding of science. There should similarly be a creation of activities to promote the public understanding of art. If we get used to seeing contemporary art, we start using our eyes and seeing the outside world in new ways. Good open assessment about art can show how a continually renewed vision of the world leads to new metaphorical forms of expression to assist us in the continuing human struggle to understand, explain and improve our lives. If we see differently, we might think differently and act differently.

There is also a need to explore further the potential for art in supporting other enterprises outside the immediate arts world. Why shouldn't artists join decision-makers, government departments, branches of industry and other work areas to propose a different way of addressing issues and problems, a way less dependent on rational discourse, more aware of what lies beneath the surface appearance of things?

Art with science

In Britain there is a long-established tradition of placing 'artists in residence' in different work settings and these have been well documented by the funding bodies over the years. These schemes are not without their problems, particularly when the artists, used to operating autonomously, have to adapt to institutional protocols and when the people in those institutions are not clear why the artist is there in the first place, but there are sometimes benefits for both parties. As scientific ideas have continued to filter into almost every aspect of our lives, it is not surprising that artists have sought out placements with science organisations and have discovered individual scientists interested in working with them. This mutual curiosity became particularly evident in the 1990s when a small number of British-based organisations simultaneously set up new initiatives.

The Arts Catalyst was established by physics-trained arts administrator Nicola Triscott in 1993 simply to explore whether such collaborations could result in interesting work. In one, choreographer Nikky Smedley engaged in research at the Theoretical Physics Department at Imperial College, out of which came a dance piece in which 'a positron-electron pair danced a duet of farewell, particles made tracks in the cloud chamber and an elongated lady danced the double slit experiment'. In retrospect this work might seem to be merely representational, but it was of phenomena which cannot properly be represented and that is an achievement in itself. It brought to non-science audiences an indication that quantum physics was indeterminate, and contemporary dance was a particularly good medium for this as it represents ideas and feeling through abstract movement conveyed through the concrete limitations of the human body. It did not anthropomorphise non-human events but, since dancers are human, it reminded viewers of what is a conundrum in the area of quantum physics it took as its subject: that the human observer may influence the event observed.

One of the Arts Catalyst's first major successes came with a commission for the artist Helen Chadwick to undertake a residency to research embryology at the In Vitro Fertilisation Unit at King's College Hospital. This resulted in her highly acclaimed (though sadly posthumous) work *Unnatural Selection* for the Arts Catalyst's 1996 exhibition *Body Visual,* in which human embryos were photographed and displayed like delicate jewels.[6] This will be analysed in more depth in Chapter 8 but it is worth noting that while the event baffled some scientist viewers, it moved and enchanted the medical scientists actually involved in the process, as it did the donors who had successfully undergone IVF treatment.[7] The Arts Catalyst has gone on to become one of the country's leading art-science agencies and its strength lies in its commitment to proper facilitation, particularly in its sensitive awareness of how the science world operates. It has good relations with Imperial College of Science, Technology and Medicine, the Natural History Museum, the Science Museum and the Royal Institution and a developing relationship with Jodrell Bank, the European Space Agency and NASA. Its project with artist James Acord and his transformation of radioactive materials at Imperial College is examined in Chapter 9.

Another new organisation, Bristol-based Interalia, under the direction of artist and physicist Richard Bright, was launched in 1990 with a series of quiet intelligent encounters where leading international artists and scientists discussed subjects to which each could bring a distinctive viewpoint. One, for example, was called 'Visions of Light' and involved international artists James Turrell and Liliane Lijn, theologian Denys Turner, and physicist Michael Berry. Another was called 'Order, Chaos and Creativity' and included cosmologist John Barrow, philosopher Margaret Boden, art historian Martin Kemp, science writer Danah Zohar, artist Andy Goldsworthy and biologist Mae-Wan Ho. The Interalia Centre was established in 1996 as the first national centre devoted to exploring the relationship between the arts and sciences, organising public meetings, business seminars (including a popular series of talks for staff at Marks and Spencer's headquarters) and 'scientist in residence' programmes at some British art colleges.

The Laboratory at the Ruskin School of Drawing and Fine Art at the University of Oxford was founded by established curator Paul Bonaventura with a view to taking advantage of the proximity of various other departments within the University. The Laboratory facilitates fellowships and related opportunities for artists to work with experts from the worlds of science and the humanities in order to generate original creative work. In 1995, for example, the artist Stefan Gec spent time with lecturers and technicians at the School of Geography to investigate the suitability of using global positioning systems to locate an artwork which tours the high seas. Using materials from decommissioned Soviet submarines abandoned near Tyneside, where he once studied, Ukrainian-born Gec built a fully workable maritime buoy. Initially *Buoy* took up positions in Belfast Lough and Dublin Bay and is currently returning to Russia, specifically Murmansk, graveyard for the obsolete Soviet nuclear fleet. The science involved may simply be an exploitation of a technological process. Artists have always loved new materials and technologies and this passion enables them to find a great deal of common ground with applied scientists, although the latter may at first be bemused by the ideas with which they are presented. But why shouldn't artists make unusual demands on material normally applied more functionally? The scientists involved have often enjoyed sharing in whimsical imaginative thinking and have naturally

Stefan Gec, *Buoy*, 1996. Positioned in Belfast Lough, Northern Ireland. Courtesy of the artist. Photo: Stefan Gec.

been happy to exploit and extend the properties of the materials which they may even have come to regard with something like affection.

One of the highest-profile initiatives in the art-science field came from a science institution. The Wellcome Trust introduced its Sciart award scheme in 1996. Led by Laurence Smaje and Ken Arnold of the Medicine, Society and History Division, the scheme was intended to supplement existing support for activities that promote the historical and cultural contexts of medicine, including the work of the Wellcome's increasingly adventurous art-science gallery, Two 10 Gallery, on Euston Road, London. The scheme was advertised, primarily to scientists in biomedicine, for projects in which artists and scientists would work collaboratively. A surprising 240 applications emerged and six awards were made. One award-winning project, for example, involved artists Heather Ackroyd and Dan Harvey who worked with geneticists Howard Thomas and Danny Thorogood from the Institute of Grassland and Environmental Research at Aberystwyth to produce large-scale 'photographic photosyntheses' in which the images were fixed by exploiting the photosynthetic properties of 'stay-green' grass through novel genetic lines. This was another collaboration which might appear at first sight to be merely about an exploitation of a new technology, although the new medium makes the artworks particularly arresting and the scientists spoke almost like artists, with a passion for the aesthetic seeming equal to that for the function of the medium. 'As you may imagine, I am inordinately fond of the colour green,' were Howard Thomas's words.

Ackroyd and Harvey, *Mother and Child*, 1998. A 'photographic photosynthesis'. Clay, stay-green grass seed, hessian, water and projector, 183 x 122 cm. From the exhibition, *Out of Sight: Imaging/Imagining Science*, Santa Barbara Museum of Art, USA. Courtesy of the artists and Santa Barbara Museum of Art, USA.

In another project, artist Stephen Farthing of the Ruskin School and surgeon Anthony Rowsell of Guy's Hospital developed a pioneering technique in cleft-palate surgery and facial reconstruction, assimilating their completely different understandings of the nature of appearance. One of the most important aspects of this project, which used digital imaging techniques, was the fact that it gave the patients involved the licence to participate fully in creating their personal images. The project inspired Stephen Farthing to assemble a digitised collection of all sorts of faces seen from a range of different perspectives, a new and developing artwork in its own terms.

A new Wellcome scheme, Science on Stage and Screen, ran in 1998, eliciting 192 applications in all (89 for the stage category alone), with three theatre awards: the theatre group Forkbeard Fantasy created a performance, animation and film show, *The Brain;* playwright Tom McGrath wrote a play about the use of viral vectors in genetic engineering; and Paul Jepson at West Yorkshire Playhouse devised a play for community touring about epilepsy, examining Dostoevsky's experience of the disease. Sciart '98 followed later in the year with another 200 applications and six more awards, among them a self-portrait in body smells developed by artist Clara Ursitti with Scottish olfactory scientist Dr George Dodd, who was eloquent about the under-representation of smell as a medium in art.[8]

Art-science collaborations have continued to increase all over the country and it would be impossible to list them all in this book. The Gulbenkian Foundation's programme *Two Cultures* was established in 1997 with the aim of encouraging artists to engage with new thinking and practice in science and technology, and the number of applications has increased exponentially over the subsequent years. The Royal National Theatre, in collaboration with the small-scale theatre group Y Touring and King's College London, have brought together leading playwrights with prestigious scientists in an intensive programme of workshops to produce one-act plays inspired by the human implications of new genetics. The trio of musicians Polly and the Phonics have planned new music inspired by the compositions of the medieval nun-composer Hildegard of Bingen, incorporating research undertaken with Ian Stewart, science writer and Professor of Mathematics at the University of Warwick, into the mathematics of game theory and also of the Fibonacci number series, which underlies spiral growth patterns in nature. Dance and theatre director Geoff Moore has undertaken research with curators in the zoology, botany, geology and biodiversity departments at the National Museums and Galleries of Wales to devise a performance piece about Darwin's contemporary pioneer in evolution theory, Alfred Russel Wallace. The Public Art Commission in Scotland has organised a residency for artist Marcus Taylor with Dr Graham Crowder and Dr Mohamed Taghizadeh at the Department of Physics at Heriot-Watt University, Edinburgh, to make a piece of public art on a rocky outcrop on the Pentland Hills to act as a data collection point, incorporating into the artwork weather instruments and indicators to measure temperature, light levels and wind direction and speed. Visible from Edinburgh Castle, the artwork takes the form of a gently shimmering disc of light which mimics the phases of the moon. Wysing Arts Centre, near Cambridge, has organised two artists' residencies in science institutions. Neal White has been based at the Human Genome Mapping Project Resource Centre (this project is described in detail in Chapter 9) and artist Natalie Taylor has worked with amenity horticulturalist Jill Raggett of Writtle College, Chelmsford, on an installation exploring the diversification of the tomato plant through selective breeding and genetic modification. In the exhibition greenhouse the energy of the growing plants was monitored and transformed to produce visual and audio feedback through devices that triggered water jets, making apparent the dynamism inherent in nature. Camden Arts Centre commissioned Simon Starling to create a flying 'butterfly chair', working with Chris Welch at the Department of Aeronautics at Kingston University, and Annie Cattrell to work with the Cognition Unit at the Medical Research Council in Cambridge to research brain scans for communication webs to be recreated in delicate glass, paper and plastic. The Hayward Gallery has commissioned four artists to make new work for its exhibition *Know Thyself: The art and science of the human body from Leonardo to now.*[9] In a project involving particle physics, Professor Ken McMullen with his colleague Michael Benson of the London Institute has organised a residency for eight leading artists including Anish Kapoor, Richard Deacon, Sean Scully, Roger Ackling and Tim O'Reilly to spend two months research at CERN in Switzerland, the site of Europe's largest particle accelerator. Artist Alexa Wright is working with Alf Linney at the Department of Medical Physics at University College, London, to develop both scientific research and artworks which challenge concepts of normality in the human face.

The ideas keep emerging and this brief list gives only a selection of the new interest, for it seems suddenly that everywhere one turns there are artists

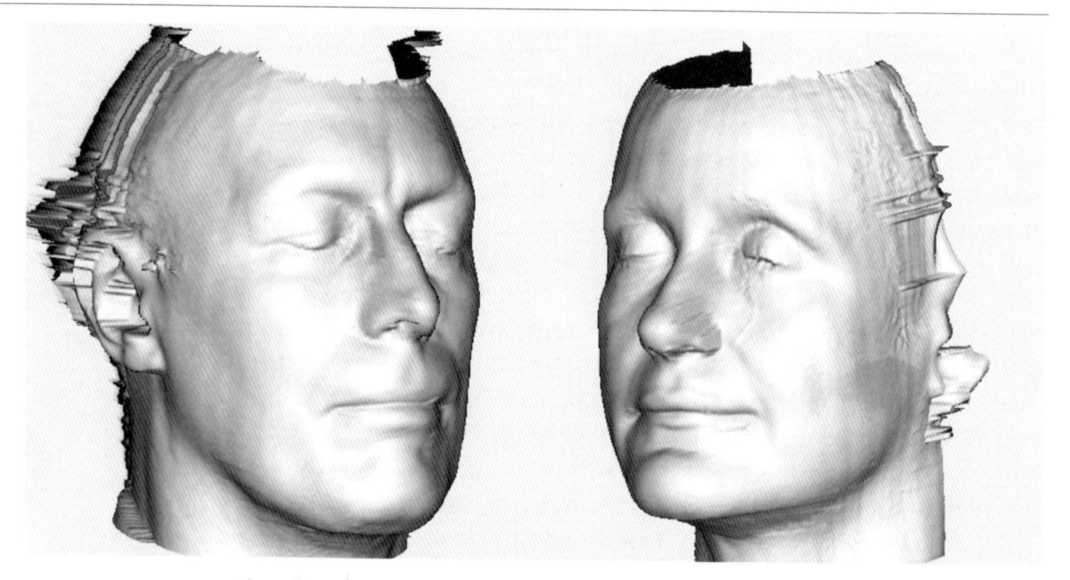

Three-dimensional scans of medical physicist Alf Linney and artist Alexa Wright, 1999, made in preparation for a project which explores the nature of normality in the human face. Courtesy of Alexa Wright and Alf Linney.

who have discovered some aspect of science and scientists who are eager to work with them. The benefits for the artists – and, as a consequence, for their public – have been wide-ranging. Scientists have been pleased to see their work given wider exposure but is there anything in the engagement which they might find directly relevant to their research?

Scientists tend not to believe that artists can teach them anything new about their own discipline, its theoretical underpinnings or wider scientific context. However, in contemporary science, dealing as so much of it does with phenomena that are 'more distant than stars and nearer than the eye',[10] we do not see directly. Visual information in laboratories is conveyed by technological media – through microscopes, telescopes, various scanning instruments – and how you set up the instruments and how you interpret the results may well be open to opinion. Artists are used not only to perceiving shapes and structures quickly, a skill which, after all, anyone might acquire with training and experience, but they are accustomed to see more than meets the eye, to see beyond a focused or perspectival viewpoint, to look 'offscreen', on the edges or even in the imagination. They are used to recognising that 'pictures' have depth, layers and multiple meanings because conceptual thinking plays as much, or more, of a part in seeing as simple visualising – as experts in experimental psychology and consciousness studies know. Suggesting alternative interpretations is the artist's natural way of working. Richard Dawkins has said: 'We mustn't forget that it is precisely the use of symbolic intuition to uncover genuine patterns of resemblance that leads scientists to their greatest endeavours.'[11] Could artists assist with this? Many possess what Martin Kemp calls 'structural intuition', famously evident in Leonardo da Vinci who sketched out the shapes and structures of physical forms and behaviours long before they were recognised by scientists and engineers.

When in 1985 the chemist Sir Harry Kroto and his team were investigating the structure of their newly discovered third carbon molecule they needed to see what it looked like. After trying out various models Kroto remembered his youthful interest in architecture and recalled the polyhedral geodesic domes of the American designer Richard Buckminster Fuller. This gave him the clue for visualising the structure of the molecules to which he subsequently gave the name 'fullerenes' or 'buckyballs'.

Theoretical physicists Chris Isham and Konstantina Savvidou have been working on a new approach to quantum theory in which the concept of time appears in two quite different ways (the time of 'being' and the time of 'becoming'). During conversations with artist John Latham they were amazed to find that he had much earlier arrived at a similar idea of time in the context of his rollerblind diagrams. Artist Mark Curtis has even gone so far as to propose a striking alternative structure for the double helix of DNA after returning to what he calls 'geometric first principles' and making careful scale drawings in the manner of Renaissance perspectival artists such as Paolo

Uccello. Although the Crick and Watson model prevails for most scientists because of the strong evidence afforded by X-ray crystallography, some are open-minded and prepared to recognise that an artist's acute visual sensibilities may make a contribution to a greater understanding of the structure of natural phenomena. 'It would be quite ridiculous not to consider something reasonable like Curtis's model,' says Robert Root-Bernstein, a physiologist at Michigan State University, 'it has novel properties that we need to understand generally, whether we're biologists or architects.'[12] Artist Adam Lowe has plans to work with American cancer researcher Merill Garnett, using his skills with digital photography to manipulate an electrical discharge into DNA procollagen to expose and auto-record the point at which cancer cells make the transition into mature differentiated forms. 'Forming and reforming the vast amount of data held within an image is an activity common to both cultures', Adam Lowe writes. 'The new artist like a good scientist relishes and interprets the infinite potential of mediated and transformed information – both visual and other.'[13]

In extreme cases the artist may become a full-time member of the scientific research team. French choreographer Kitsou Dubois was first attracted to explore the sensation of moving in zero gravity because it was a wonderful medium for 'experiencing a perfect state of dance ... corresponding to those magic moments which in dance one is always trying to recapture'.[14] However, as she became involved with the French Space Agency she began to develop dance-based protocols for astronauts training in parabolic flight,

Kitsou Dubois, *Gravity Zero*, 1999. The dancer's direct experience of moving in zero gravity draws together the different aspects of research conducted by individual members of the Biodynamics Group at Imperial College. The video of her parabolic flights is also shown in galleries as an artwork in its own right. Courtesy of the artist, Ki Productions and the Arts Catalyst, London.

assisting them in combating the debilitation of space sickness. In a project facilitated by the Arts Catalyst she went on to set up a new interdisciplinary science-art collaboration with the Biodynamics Group at Imperial College. As a measure of its credibility in the science world, the Project Team has had a research proposal accepted by the European Space Agency. In this, physiological evidence is provided to support the choreographer's contribution towards helping astronauts adjust to weightlessness. Kitsou Dubois's work as art has been displayed on video which captures the sheer joy of moving in an unnatural environment. On one level it communicates the age-old human aspiration to move free of any earthly encumbrance, but because of the scientific context, it simultaneously inspires the viewer to think about the reality of space exploration. Kitsou Dubois plans to be the first artist to go into orbit when the International Space Station has been constructed.

Not all new art-science endeavours will go this far, however, and many seem still to be in the experimental stage. They provide for artists and scientists alike an exhilarating but also a frustrating challenge. For despite the commitment of participants and funding bodies, many current art-science collaborations are only partly successful. There are still deep conceptual misunderstandings on either side which need to be properly aired, as much to do with cultural differences as with the actual material under artistic or scientific scrutiny. In some collaborations there is a clear sense that the artist is lucky to be allowed into the laboratory to indulge in incomprehensible behaviour, all to produce a piece of work which, to put it kindly, may eventually be regarded as puzzling, while all around busy scientists are getting on with terribly important research and 'real' work. As things stand, the art in many art-science collaborations is the icing on the cake – decorative, eye-catching and of superficial educational value. There are wasted opportunities on both sides.

In a well-thought-through collaboration some form of cultural induction may be necessary. In the first place each party needs to learn a great deal simply about the other's practice. Artists who may have thought that scientists strut around in white coats methodically measuring data all day may be interested in discovering how often they have brainstorming meetings, formal and informal, to discuss unpredictable results. A tangible sense of excitement prevails in some research facilities and, contrary to general belief, scientists often discuss and argue about the ethical and practical implications of their work. Scientists who believe that artists dash in, make direct contact with the Muse and instinctively produce a lovely painting, may be surprised to discover how hard they work, discarding drafts and sketches, trying out ideas, altering, adjusting and working for many hours at a time on a tiny detail. Art is no more a part-time or amateur pursuit than science is. And scientists who believe they could make a career in both simultaneously should remember how many years they have taken to learn the discipline of their own culture and how time-consuming and absorbing the proper practice of science is. Art looks easier than it is, and requires equal dedication. Moreover, the complete freedom enjoyed by artists to make what they wish of any subject is a curse as much as an opportunity. Solely responsible for their work, they can fail miserably and have to harden themselves in the face of flippant or casual judgements. Scientists have to get used to a highly charged competitive ethos but they work in teams and they are less vulnerable as individuals, even though they face the risk that years of research may be rendered useless if another team publishes its results first. This mutual learning experience may produce little in the way of 'work' but it can subtly alter perceptions and enable both sides to understand each other's field better. This may have beneficial repercussions particularly when it comes to understanding the controversies associated with much scientific research.

In order for new collaborations of any sort to occur, however, new opportunities need to be made and supported. Indeed, the education and training available to both artists and scientists might need some reappraisal to allow such joint schemes to happen. An increasing number of art schools run courses in 'contextual studies' or 'visual studies' instead of simply 'fine art'. The Contextual Practice Network, for example, is a confederation of art schools which offer fine art courses concerned with encouraging artists to work in public and social contexts.[15] More and more professional art galleries and agencies are organising events, courses and residencies for artists in science and there needs to be further encouragement for arts organisations to become more aware of scientific issues in programming, commissioning work and in supplementary activities.

Science institutions are also beginning to take the initiative. Imperial College hosts the orchestra Sinfonia 21 and has a regular programme of artists' residencies, exhibitions and talks. The National Institute for Medical Research at Mill Hill, London, has a modest visual arts programme which explores the potential for artists and craftspeople to create new work to enhance the physical site, with a proposed regular programme of artists' residencies, art-science collaborations, arts activities, lectures for staff and a related education programme for the wider public. It would be simplistic to claim that the presence of art merely lightens and brightens the scholarly atmosphere in science-dominated places, for it is to be hoped that the different ways of looking at all aspects of the world might also make their inhabitants recognise that there are many ways of thinking profoundly.

This last aim may be hard for scientists to understand, especially when all they want is an attractive artwork which can be put on display to decorate a sombre workspace or to illustrate the science and hopefully elucidate its mysteries for the general public. The patterns formed by groups of neurons, or galaxies, for example, could easily be reproduced to make pleasing pictures for galleries. But how far can this be regarded as art? Science illustrators abound and are highly skilled at making representations of phenomena which may be said to have an 'aesthetic' quality. But, devoid of individual interpretation or passion, this is illustration, not art. Artists, like most of the public, would find it inconceivable, for example, that anyone could talk about the beauty of cellular structures if it transpired that the cells in question were malignant. No matter how attractive the molecular shapes appear – demonstrating pleasing symmetries or harmonious colours – the fact that they constitute cancer cells evokes a chill that dispels all real ideas of beauty.

The nature of any artwork resulting from a collaboration may challenge our ideas of both science and art. One which ends up in a gallery or other exhibition space, or even on-site at a laboratory or university, for example, may reflect little of its scientific genesis and may provide the viewer with meanings which are obscure in the rational sense, although rich with implication in the subliminal, abstract or emotional sense. We should not always anticipate a straightforward scientific explanation, although good explanatory material should always be available nearby so that viewers can make their own associations. On the other hand, some artworks may be enriched because they are underpinned by an acknowledged intellectual understanding and explanation of the science involved. Just as one can read a novel and switch back and forth between a rational understanding of an issue and an emotional response to the human circumstances surrounding it, so one can respond to an artwork on a basis which is both sensuous and intellectual. New types of artists may emerge, making a virtue of their inherent urge to dissent by bringing a healthy questioning heterodoxy into other work or social environments.

Many scientists would welcome this interest and would be willing to open the doors of their laboratories and spend time explaining what they do, but they should recognise that artists are not there simply to record or interpret givens and cannot be regarded as mere image-makers. If artists are let in at the door, they may ask questions. Why pursue this area of research rather than that? Why are we narrowing our focus to examine this one small phenomenon without taking into consideration its place in a wider context? Do we need to know this at all? Most scientists would not welcome these presumptuous questions or even see their relevance. Surely the aims and objectives, the moral considerations connected with science, should be dealt with by the research councils, the ethics commissions, the government committees, the think-tanks, the pressure groups and the lawyers? The place for a dissenting viewpoint is outside the laboratory.

But the potential for profound levels of engagement should be acknowledged. If artists become well informed in science they are bound to want to contribute. They inhabit a peopled universe and even if we *are* on a 'darkling plain' where there is no single comforting belief system and where suffering predominates, perhaps art's greatest value lies in its ability to assure us that we are not separate or alone on earth. 'Humankind is central in my work,' writes German photographer Andreas Gursky, 'even when, in exceptional cases, it can be reduced to the point of invisibility.'[16] Drawing more on intuition than received knowledge, artists always start from the human perspective. The personal, the 'I', which features in a great deal of contemporary art, poetry, novels and films, in the best work is not a sign of narcissism. We identify with each other and we are reassured by recognising our own feelings and thoughts in the expressions of others. Science is advancing rapidly, presenting a challenge to our world-view, so that it is imperative that artists help us make sense of it – literally and metaphorically. When science bounds forward too dramatically, too fast, artists may take it as their responsibility to remind us through their work that we are not demigods poised to control the universe. We are humans here and now, still encumbered by our primitive bodies and still controlled by our contrary passions. We need to communicate with each other in an isolating world. We need to keep our sense of perspective.

Art and science are two very different cultures. Within the human mind they present different forms of understanding. They are not directly

Eduardo Paolozzi, *Newton*, 1988. Bronze, 44 x 33 x 60 cm. Courtesy of the Tate Gallery, London. © Eduardo Paolozzi, 1999. All rights reserved DACS.

Drawing on Blake's original image (see page 34), the Scottish-Italian artist Eduardo Paolozzi presents a contemporary view of Newton in this sculpture (a much larger version stands outside the British Library in London). As in the Blake painting, Newton's pose is crouched and he is still preoccupied with measurements, but his eyes are visionary, inspired by the eyes of Michelangelo's *David*, the subject of Paolozzi's first exercise as a fine-art student at Edinburgh College of Art. For another work Paolozzi has quoted Einstein: 'Knowledge is wonderful but imagination is even better.'

comparable and they are often incompatible. The human mind is equipped to draw on both systems of knowledge and we need both to survive. We think and order and distinguish; we also feel and guess and connect. This dual nature is not limited to the workings of the mind, but relates, too, to the workings of the social world we live in. Let the differences remain, but each culture should respect the other. And artists may find that, after all, they have a *function* which can be genuinely valued, engaging with science in the making of images for the better understanding of ourselves.

Notes and references

1 Jo Shapcott, 'Leonardo and the Vortex', from *Phrase Book* (Oxford, Oxford University Press, 1992).

2 Quotations in italics are as follows, in order: T.S. Eliot, 'The Waste Land'; Robert Graves, 'The White Goddess'; Purcell, *Dido and Aeneas;* W.B. Yeats, 'Sailing to Byzantium'.

3 This state of affairs has become compounded as art schools have strengthened their affiliations with universities. Students are assessed on the basis of academic achievement and it is difficult to place practical studio work within this context. The Research Assessment Exercise, which forms the basis for assessing the research status of university departments, is based on the amount of original scholarship they generate. Written research is much easier to account for than practical artists' research.

4 Rick Gibson clearly did get stopped; see chapter 2, note 3. The controversy over Robert Mapplethorpe's photographs in the USA is also a sign that some sectors of the public want to draw the line somewhere.

5 'I am disappointed by that stroke of death, which has eclipsed the gaiety of nations and impoverished the public stock of harmless pleasure.' Dr Johnson, on the death of the popular actor Garrick, *Lives of the English Poets* (1779–81). It would be too trite to claim, although I'm sure some evolutionary psychologists might, that art is merely a display of sexuality and power – young cubs rout old wolves.

6 Sadly, Helen Chadwick died before the exhibition *Body Visual* opened in May 1996, at the Barbican in London; it had a subsequent tour of hospitals and galleries (1996–8). The exhibition, which also featured the work of artists Mitzi Galli and Donald Rodney, was managed by the Arts Catalyst; catalogue from the Arts Catalyst.

7 Helen Chadwick used 'spare' embryos in her work. Donors signed a consent form, drawn up with advice from the Human Fertilisation and Embryology Authority, which gave the artist permission to use these in a non-living state.

8 Information about Sciart 1 and 2, from the Wellcome Trust, London, where the scheme is administered. The Sciart scheme was subsequently made independent and funded by the Wellcome Trust with the Arts Councils of England and Scotland, the British Council, and the Calouste Gulbenkian Foundation, and sponsored by the National Endowment for Science, Technology and the Arts (NESTA).

9 Exhibition at the Hayward Gallery, London, Autumn 2000, curated by Martin Kemp and Marina Wallace.

10 T.S. Eliot, 'Marina' (London, Faber, 1930).

11 Richard Dawkins, *Unweaving the Rainbow* (London, Allen Lane, The Penguin Press, 1998), p. 186.

12 For fuller information on Curtis's DNA model see Ayala Ochert, 'Deconstructing DNA', in *New Scientist*, 16 May 1998.

13 Adam Lowe, artist, in correspondence with the author, 1998–9.

14 Kitsou Dubois, quoted in the Arts Catalyst's exhibition of her video work at The Lux, London, Spring 1999.

15 The co-ordinator of the Contextual Practice Network is Judith Rugg of Exeter School of Art, University of Plymouth.

16 Andreas Gursky, quoted in the leaflet accompanying an exhibition of his work at the Serpentine Gallery, London, in January 1999.

4

Between Explanation and Inspiration: Images in Science

Ken Arnold

There is no science without fancy, and no art without facts.

VLADIMIR NABOKOV, 1966.[1]

Introduction: scientific images and works of art

Two images: some similarities and some differences. Both strike us as abstract pictures, etched in high-contrast, vibrant colours; both evoke a sense of fantastic landscapes awaiting human exploration. More prosaically, both are fundamentally composed of bold structures and intricate patterns. But differences abound too. The first, it turns out, shows the retina in a guinea pig's eye; the second, crystals of a cadmium compound. The first is a colour-enhanced scanning electron micrograph; the second a photograph. The first, and this is probably the most important difference, was taken by David Furness of the Department of Communication and Neuroscience, Keele University; the second by Felice Frankel, artist in residence at the Massachusetts Institute of Technology. The first then would seem to be a scientific image, the second a work of art.

As we look at the two images we entertain the notion that both science and art are essentially ordering activities, part of the universal human inclination to find, expose and celebrate the world's structures and patterns. Even more fundamentally, they gesture towards the fact that both art and science are expressions of a common intellectual curiosity – the profound human desire to know things, which often starts with the possibility of envisioning and therefore of making a picture of them. But whereas scientists need to account for their illustrations, either explaining and thereby weaving their significance back into the experimental and mathematical logic of their argument, or alternatively, simply presenting them as incidental by-products of a strict methodology which places little or no emphasis on images, artists tend instead to insist that the visual products they create stand or fall on their own.

Guinea pig's retina. Winning image from Biochemical Awards, 1998. Scanning Electron Micrograph. Courtesy of Dr David Furness, Wellcome Trust Medical Photographic Library.

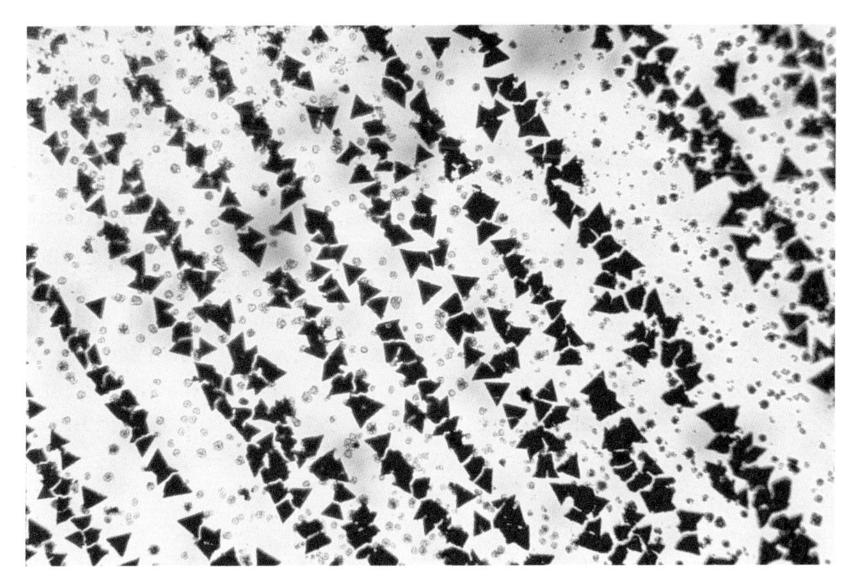

Felice Frankel, *Nucleation of Crystal Growth*, photograph of assembled nanoclusters of cadmium selenide crystals, from *On the Surface of Things*, 1997. © Felice Frankel.

No doubt, images in both science and art have an extraordinary capacity directly to arouse our attention, but only in the former are they judged by their ability unambiguously to communicate a specific message. Or as Martin Kemp has suggestively put it, whereas scientists aim for a 'verbally controlled explanation of the processes' they are investigating, artists instead insist that the sum of a work's effects 'must ultimately be its end'.[2]

Visual similarities between our opening two pictures juxtaposed with the seemingly enormous differences in their cultural significance and the intentions of their makers, provide a point of departure for this chapter which looks at scientific images and their cultural impact. It is worth recalling in passing that more than 20 years ago the philosopher and historian of science Thomas Kuhn made the important point that the comparison of art and science needs to be made at three levels: products, activities and public response.[3] Most of this chapter will be primarily focused on the first two, though it should be remembered throughout that though the general public seems equally estranged from both contemporary art and science, the latter tends to be produced for other scientists, whereas only the most nihilistic of artists would say the same of their own work. This chapter, then, starts by outlining the nature and history of the visual parts of science and then moves on to consider the impact of various aspects of technology on what might be termed 'visual practices' within science. Next, it takes a closer look at the aesthetic legacy of all this 'visual science' and especially at its appeal for artists and viewers. Finally, in order to complete the circle, it will briefly speculate about the possible fruits of a renewed involvement of artists within scientific image-making. We thus begin with the aesthetics of science and end with the empiricism of art.

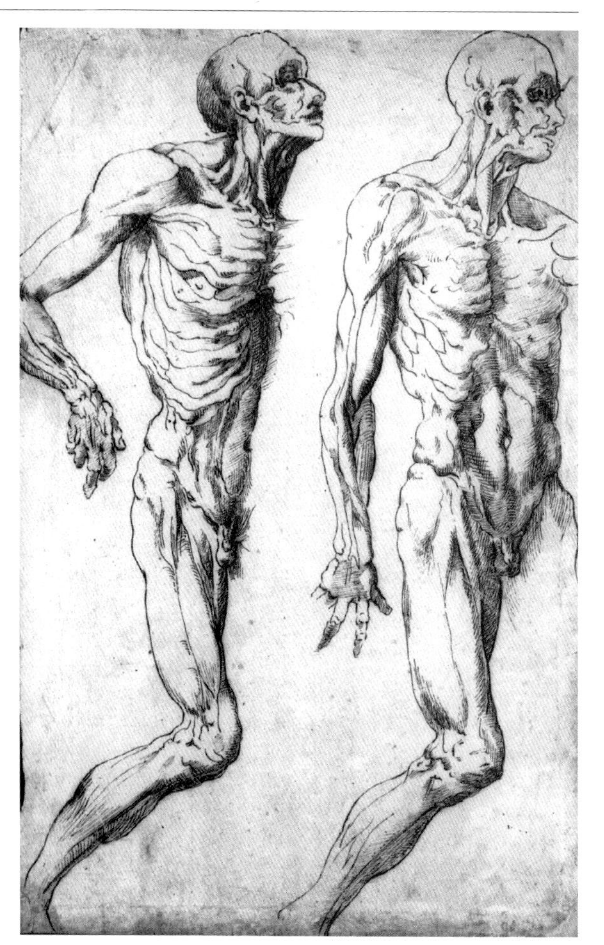

Pen and ink anatomical drawing attributed to circle of Bartolomeo Passarotti (1529–92). Wellcome Library, London.

Visual thinking: the art of science

The beginnings of modern science in the Renaissance partly rest on the emergence of a range of inquiries defined by visual practice. To be sure, much of the import of the work of Copernicus, Galileo, Harvey, Newton and their fellow pioneering natural philosophers was attached less to the pictures and more to the logic of their arguments, the validity of their factual evidence and the precision of their geometry and mathematics. But the new type of publicly accessible, factual knowledge that they sought to create did nonetheless have a strong visual component, particularly in the areas of geometrical perspective, the more visual parts of astronomy, anatomy and various parts of natural history.

The Renaissance saw the emergence of the cluster of sciences which are commonly grouped under the term 'natural history'– the observational and descriptive parts of botany, entomology and various parts of modern geology, zoology, and ornithology. In part, this loose collection of pursuits was held together by a general, if vague, theological commitment to the idea that studying nature could be viewed as a largely 'outdoor' form of religion in which 'reading' the book of nature somehow complemented an understanding of the other book, the Bible. Philosophy, as well as theology, bound the new natural historical sciences. All were based on gathering specimens to turn into vast empires of facts, and all of them made use of careful visual observation and pictorial representation.

As these traditions of 'visual science' evolved, the support that natural theology lent to them gave way, in an increasingly secularised world, to a range of more abstract arguments for the significance of the visual scrutiny on which they were based: mathematical interests in form, structure and patterns, biological interests in growth and sex as a key to classification, and, eventually, evolutionary interests in change and inter-species relations. The success of earlier visual sciences (perspective, colour theory, anatomy, natural history and bits of astronomy) in due course provided inspiration for a second family of intellectual disciplines – anthropology, ethnography, archaeology, sociology, and later psychology – all of which sought to bring positive factual knowledge to a new subject – humankind. In their early histories at least, these human sciences also placed a strong emphasis on visual evidence. To be sure, classification, inductive logic, abstract theorising and any number of other verbal and logical methodologies all played crucial roles in their formation, but much of the groundwork in establishing their foundations involved the creation of pictures, diagrams, illustrated catalogues and other visual representations.

Visual instruments: machines for seeing

The visual sciences have always been characterised by perpetual evolution with occasional revolution. The most energetic instigators of change within them have been the inventors of the many and varied scientific instruments developed to overcome the natural limitations of human eyesight. The two most significant tools employed to extend the visual reach of science were the telescope and microscope, which in the hands of pioneers like Galileo and Robert Hooke were soon shown to provide keys to previously unimagined physical phenomena just out of sight, in the realms of the very large and the very small. In considering the impact of these visual scientific instruments, it is important to acknowledge the extraordinary dexterity employed by those using them and the fact that this was matched by the precise skill of the engravers, printers and publishers, who were able to turn the scientists' dramatic personal discoveries into publicly disseminated images – images which, of course, rapidly became icons for the whole enterprise of visual science.

The increasing use of telescopes and microscopes set in train a gradual shift from the use of the naked eye to the manipulation of machines for seeing.

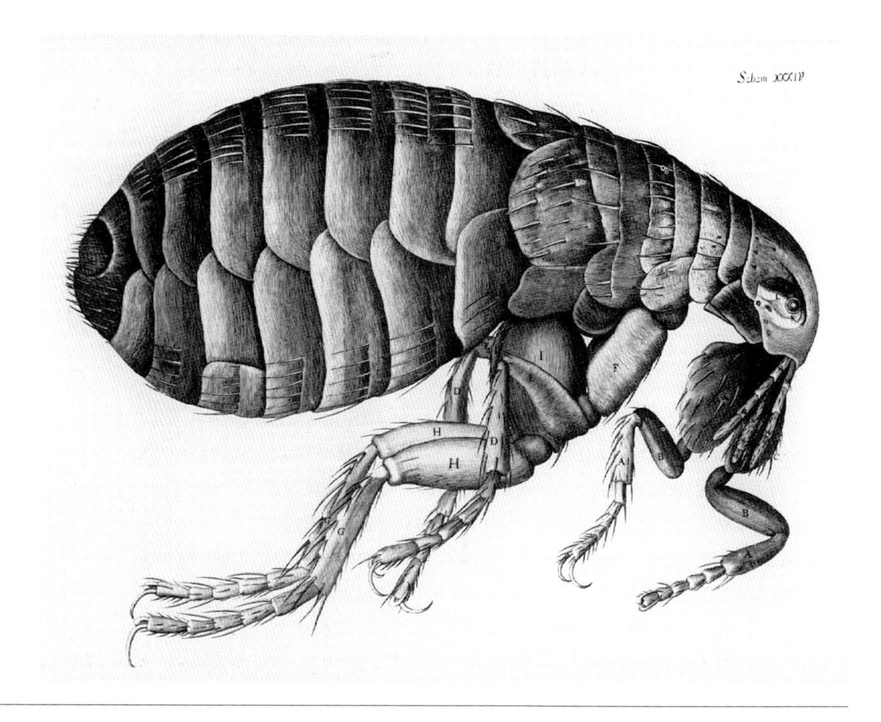

Image of a flea, from Robert Hooke's *Micrographia*, 1665. Wellcome Library, London.

One important consequence of this change was that the trust traditionally invested in the observer's senses, abilities and honesty would now need to be extended to the observers' instruments and the claims made about exactly what could be seen with them. Scepticism about whether and how to trust the instrumental extension of human sight was similarly expressed about another technical innovation – the camera – which was destined to revolutionise almost all aspects of the visual sciences (and, coincidentally, the arts). The older inventions of camera obscura and lucida did little to prepare scientists for the eventual impact which photography was to have on geology, medicine, biology, zoology, meteorology and indeed every branch of science, including physics. From the late nineteenth century onwards, the further development of photographic procedures has continued to refine methodologies for capturing and recording ever more intricate and hidden natural structures and patterns, to say nothing of objects in motion and changing over time. Photography profoundly altered the process of obtaining visual information, representing both a new type of permanent record of the observational act and a new medium through which scientific information might be communicated. But it also dramatically changed how science was done and how it was publicly perceived.

Scientific photography paved the way for the development of many new visual media – and indeed new disciplines – in the sciences. Any history of the visual aspects of science would probably have to reserve at least a chapter each for the stories of the impact of X-rays, at the end of the nineteenth century, and then, mostly in the twentieth century, of the various methods of recording the behaviour of atomic and subatomic particles. At the opposite end of the spectrum, the application of radio-technology in observational astronomy would also require detailed analysis, with yet another chapter or two to describe the explosion of even newer imaging techniques, employed with increasing audacity during the second half of the twentieth century, from electron microscopes to Nuclear Magnetic Resonance Imaging (NMRI). One can talk without hyperbole of entirely new 'visual worlds' opening up across universes, both without and within the human body, in almost every semi-decade since 1960.[4]

However, before we are carried away in considering the serious application of these breathtaking innovations – the telescope, microscope, camera and bubble-chamber – it is worth recalling some rather more frivolous examples – those visual instruments produced primarily for recreation and entertainment. These were to have quite a different kind of impact on the public imagination. Into this category come many scientific toys such as the highly elaborate orrery spheres and perspectival machines used by visual entertainers, the zootropes and visual instruments developed specifically to create illusions, as well as any number of toy microscopes, telescopes and other gadgets, displayed in their full splendour with the advent of 'popular' and sensational natural history museums. To the perpetual frustration of serious scientists, this rich and varied array of cunning apparatus used to demonstrate, often with a showmanlike panache, the more fantastical and visual aspects of science, has long provided the general public with its chief source of scientific perception. This tradition of purveying science as a tricksy visual display still lingers today in the form of rather tame chemistry sets, Christmas lectures, popular TV science programmes and some of the 'interactive' displays in science centres. For the public, science is often seen as a subject little concerned with deep concepts or abstract thinking and is more to do with dramatic visual display and sleight of hand.[5]

Visual science: the legacy

It would be impossible to provide a single history of the immense and divergent body of visual material that science has produced, let alone a unified interpretation of its significance. Nonetheless, even with a skeleton outline of its history, it *is* possible to make a number of important general points about the role of images within these scientific traditions.

Firstly, when we consider the body of these pictures as a whole, it is clear that representations of various aspects of science have always been, and indeed still are, influenced and enriched by an aesthetic input. Few now doubt that this was the case in the Renaissance, when artists and scientists could be one and the same person, and this state of affairs continued during the centuries thereafter, when training in seeing and drawing played an important part in many areas of scientific education. And while visual skills are far less of a requisite for contemporary scientists, notable exceptions persist, especially in the biological and medical sciences. Here, much of the most innovative scientific data recently produced has arrived dressed up in radically new techniques of visualising and picturing – electron microscopy, for example, crystallography and colour-enhanced NMRI images. One of the significant things about the images produced by these techniques is that the colour and definition which characterises so many of them is frequently 'artificially' added to the raw data, often at computer terminals, in a manner that strongly suggests considerations that are not purely 'scientific'. In these areas of scientific image-making, it does not seem too outlandish to assert that the absolutely strict divide between a mechanical 'recording' of what is found and an artful construction of what can, with prompting, be 'seen' is often blurred. Nature and the minds that discover it thus become intricately interlinked, and within the pattern of their intermingling one discovers aesthetic judgements at the heart of the practice of science.

The second point, however, is that the scientists who produce these often stunning pictures tend not to consider themselves as fundamentally different from other scientists. Most of them see their pictures simply as tools to further specific areas of research, in just the way that others in their field might view a graph or set of statistics. Some might concede that conventions have become established so that images are created in a particular 'style', while others will admit that their images may be open to interpretation, both of which sound like the sorts of things that are said of images in the world of art. And certainly this tendency towards the notion of an 'aesthetic' science is further highlighted by the fact that some of these images have been put on show as if they were artworks – framed and captioned in conventional art-gallery fashion. However, within the scientific communities, there is no doubt that an unbridgeable and enduring gap exists between image-makers in science and those in the arts. The aims of scientific image-makers are entirely coterminous with the rest of their professional colleagues who use different methods of presenting data. And though the images they produce are undoubtedly aesthetically attractive, they will continue to be seen by the scientists who make them as data and tools and not prized objects in themselves.

The third point that needs to be made is that the images created by 'visual scientists' never stand simply as literal registers of what they see. These pictures are better understood as diagrams containing encoded information, which, either explicitly or implicitly, draw on visual conventions adopted by both creator and user. This is important, because the scientific content of an image can be transmitted successfully only if creators and viewers share the same interpretation of what they are looking at. In fact, scientific images inevitably carry with them a series of contexts, from the technical apparatus of captions, scales and codes which guide the viewer's 'reading' and understanding of them, to the broader cultural and social contexts which constitute a framework of belief. The existence of these contexts makes for another distinction between primarily scientific and primarily aesthetic images. One can argue that any number of images can be moved from one domain to the other simply by changing the contexts in which they are presented. This, after all, is what has happened to most of the 'classic' pictures from the history of science, and what a number of recent 'art' exhibitions of scientific images have deliberately demonstrated.

The fourth point is that although the later history of many of the visual sciences has accorded less and less importance to the reproduction of visual information, many of the pictures at the heart of their early history, at least, were far from being merely 'illustrative'. As the science writer Stephen Jay Gould has put it, 'scientific illustrations are not frills or summaries, they are foci for modes of thought'.[6] Thinking and visualising are not mutually exclusive activities and the term 'visual thinking' has been coined to represent a mode of perception and understanding which is quintessentially beyond logic and words. The significance and weight of 'visual thought' can be gauged by the number of instances in science where a set of images derived from the world of sense perception has been replaced by 'artificial' visual images, and even more so by models, which go on to generate their own reality. Thus geological maps, at a macroscopic level, and models of the atom, at the opposite end of the scale, have come to exert extraordinary explanatory power precisely because they are easier to deal with than nature itself. Some might even claim that visual thinking is the most important form of understanding, so that science's crowning achievement – the formulation of explanatory theories – becomes an exercise far more akin to the process of making a picture than formulating a sentence with rational language.[7]

However, we should not forget that the use of images to understand reality co-exists with an equally strong tradition which has instead placed its emphasis on logic, measurements, numbers and quantities, all of which are presented in the aesthetically emaciated form of statistics and graphs. Thus, for example, nineteenth-century German and French laboratory-based research in organic chemistry and physiology effectively turned a variety of the body's activities into systems of measurements. In medicine in particular, but also throughout the observational sciences, a range of new instruments was introduced in the late nineteenth century which allowed individual phenomena to be automatically recorded in a form seemingly more truthful and trustworthy than traditional descriptions and nomenclatures which were seen to be bound up with the fallibility of human observers. Carl Ludwig's kymograph, which enabled the automatic translation of bodily functions into quantitative records, stands as something of an emblem for what amounted to a radically new, non-naturalistic representation of the body. The role of the observer's personal understanding, intuition and memory, so fundamental to much pre-twentieth-century scientific practice, was thus gradually replaced by the more focused technical skills of reading and interpreting figures.[8] Science, it seemed, would be stronger the more it rested on the certainty of machines and the less it was supported by the frailty of hand-eye-brain coordination in human observers. This very powerful and, some would contend, still dominant model for scientific knowledge thus dispensed with problematically ambiguous images and illustrations. In our own time, indeed, non-visual models for capturing information seem to be in the ascendant – the digital sequencing of the human genome, the science of bio-infomatics, the increasing use of digitally analysed taxonomies to replace observations in the field, computerised models of neural networks and so on.

The visual appeal of science: art in science

Despite the trend noted above, artists will be drawn to the astonishing array of scientific images which has been produced throughout the history of science. This interest has been manifested by an increasing number of publications and exhibitions which have made scientific images accessible to a wider public. More recently in the UK these have included an exhibition on the body curated by Deanna Petherbridge (*The Quick and the Dead*), which started a European tour at the Royal College of Art in 1998, with another taking the body as its subject in Autumn 2000 at the Hayward Gallery: *Know Thyself: The art and science of the human body from Leonardo to now*, curated by Martin Kemp and Marina Wallace.[9]

What is the attraction of scientific images for artists and viewers and why have we recently become increasingly drawn to such images? Of course many images have a quality which is undeniably 'aesthetic'. They are 'beautiful', whether they are regarded in their particular contexts or viewed as things apart. Scientists often express a sense of wonder and exhilaration at the forms they deal with and want to share this with others – the perfect numerical order of a Fibonacci sequence displayed in the whorls of a shell, the movement of clouds captured in film, the delicate asymmetry of cells, even the scratchy traces of subatomic particles in cloud chambers. It is possible to view these objects as lovely images in themselves and as lovely manifestations of a beautiful impersonal world; though we should not forget that scientific images can also be gruesome and freakish. For most artists, and indeed for most of the public, the world is not impersonal, and scientific images put on display will be viewed subjectively, in relation to the viewer's own feelings and experience, just as art objects are in galleries and museums. An intellectual understanding of what an exhibit actually is may inspire dispassionate awe and greater knowledge but it may also evoke an emotional response, in which a fearful curiosity might play an inordinate part because we are being shown things we are not normally equipped to see, whether they are images of distant galaxies or the microscopic detail of our own bodies. Viewing scientific images we are perhaps confronted – sometimes more vividly than we are through conventional artworks – with a sense of our own impermanence and fragility, where our consciousness of boundary and identity and scale is displaced. A recent exhibition in Germany of actual human corpses frozen, encapsulated in plastic and placed in ordinary domestic poses, drew vast crowds, eager to see objects which may be commonplace (although not in such grotesque poses) for many biomedical scientists, but which are no longer visible in our sanitised life in the West.[10] More reassuringly, artist Benedict Rubra who has examined human cells in detail in preparation for non representational paintings, has reported feeling a sense of calm as a result of his deeper appreciation of the structures of human tissue, as though the knowledge has in some way been able to dispel mystification and dread.[11] The visual material which science has until relatively recently kept to itself – whether of body parts or high definition microphotographs, pictures of the galaxy from the Hubble Telescope or traces of subatomic particles – is now becoming much more available in the public domain and our growing curiosity may say as much about our desire to view ourselves differently as it does about our desire to find out what the scientists have been doing and what it looks like.

Artists are nothing if not avid consumers of images and image-making technologies. The products of their labours have always shown evidence of creative acts of borrowing and reinvention. Not surprisingly, therefore, there is a long tradition of artistic exploitation of the extremely rich body of visual science available. From the Renaissance, if not earlier, artists have seized scientific images, their patterns, forms, colours and structures, but also the concentration of ideas about the world that they embody, as raw material to reuse in their art. In our own era, when many critics wonder if art is in danger of losing its way somewhere between the conflicting roles of decorating the world and undermining its very foundations, the seeming certainties of science lurking behind these images may have made this iconography particularly beguiling. Writing about images of cells in twentieth-century art, Maura Flannery has commented that 'it is ironic that when twentieth-century artists broke away from realism they grasped at elements of another realism: that of the microscopic level.'[12] Whatever the causes, science is apparently a subject that no contemporary artist can afford to ignore. And, as a recent *New York Times* article has described it, 'almost no scientific invention, principle, fact, technique or technology seems immune to the artistic appetite'.[13]

Many contemporary artists have been drawn to the images, methods and ideas revealed in the visual practices of biological and medical sciences, which continue to be the most visual scientific disciplines and the ones to which we can most immediately relate. Just a few examples serve to make the point: Mark Francis's large canvases that bring together art and science perspectives on

Simon Robertshaw, *The Errors of Art and Nature*, 1994–7. Video installation, X-rays. Courtesy of the artist.

The installation counterpoints two videos: the Eugenics Society film *Heredity in Man*, which traces the heredity of a family 'of good stock' with the aim of showing how the human race can be improved by preventing people from 'undesirable groups' from breeding, and a video made during the artist's residency at a psychiatric hospital where he worked with young offenders.

issues of creativity and union in patterns that suggest swimming sperm; Helen Storey's audacious haute couture dresses that parade the biological phenomenon of the 'primitive streak'; Simon Robertshaw's installation *The Errors of Art and Nature*, which dwelt on such troubling aspects of science as eugenics and the psychiatric treatment of young offenders; David Jane's work dealing with encephalitis; two different series of paintings by Paul Harrison and Zara Matthews, both of which relate to images of chromosomes and nerve cells; Adam Gray's work which, by presenting human tragedies such as shipwrecks as almost insignificant details in vast calm canvases, explores the emotional as well as technical aspects of scale – that ubiquitous coordinate of contemporary scientific enterprises; and finally Anna Hill's piece *Still Like Dust I Rise*, which powerfully juxtaposes material relating to biological science and the emotional response to death, as well as a newer piece *View from Within*, which comprises enormous room-filling images of the inside of the retina.

The visual response of artists: an aesthetic feedback?

The artworks described above take science as a starting point but they are artworks in their own right, not illustrations of science. However, in some recent art and science collaborations, there has been an unspoken implication that the function of the art is merely to serve as a user-friendly filter for science, which might otherwise be too difficult for the uninitiated to deal with. There is nothing in itself wrong with work which seeks to fulfil this function, nor with this didactic philosophy for science education. However, there is a danger that the art can end up being merely a servant to the 'higher' ends of science. Much of the work produced within these confines is forced to do little more than bear decorative witness to the undeniably extraordinary patterns and structures in the world which science continues to reveal. In the worst examples of such work, the science becomes diluted and the art serves only to illustrate. However, science and art can make much more out of their shared interest in image-making.

Historians of both art and science have uncovered many instances where artists have done far more than merely illustrate and make accessible concepts already established in science. Such claims have, for example, been made about Constable's cloud studies as a branch of the science of meteorology and Turner's insights into the properties of light and atmosphere.[14] Much too has been made of conceptual links between Cubism and Einsteinian relativism, though here the influence between the two is more convincingly described as operating at the level of *zeitgeist* rather than in any directly causal way.[15] It is presumably also at this level that we should take the recent suggestion made by the Australian physicist Richard Taylor that the drip paintings of Jackson Pollock provide notable examples of chaos theory.[16] For as Martin Kemp has pointed out, it is perfectly possible for artists to prefigure in their work physically significant ideas which are only later fully explored by science.[17] Kemp's own researches into this area have revealed dozens of instances (some better known than others) of artists working in ways that shed much light either on specific scientific subjects or more generally on the scientific method. In this vein he has made clear the scientific dimensions of a variety of both historical and contemporary artistic enterprises: Berenice Abbott's photographs that caught aspects of motion, like wave interference patterns, too rapid or subtle for detection by the human eye, and the deft experimentalism that is clearly employed by Joseph Albers in his abstract colour paintings, which bring out the subjective nature and fluidity of colour perception.[18]

All this evidence points to something much more profound than mere illustration. Such sparks of cross-disciplinary inspiration seem, however, to require some sort of neutral territory – an area of interest not completely monopolised by either domain. In short, artists and scientists seem to have identified visual images, and, maybe more specifically, the technologies and sites

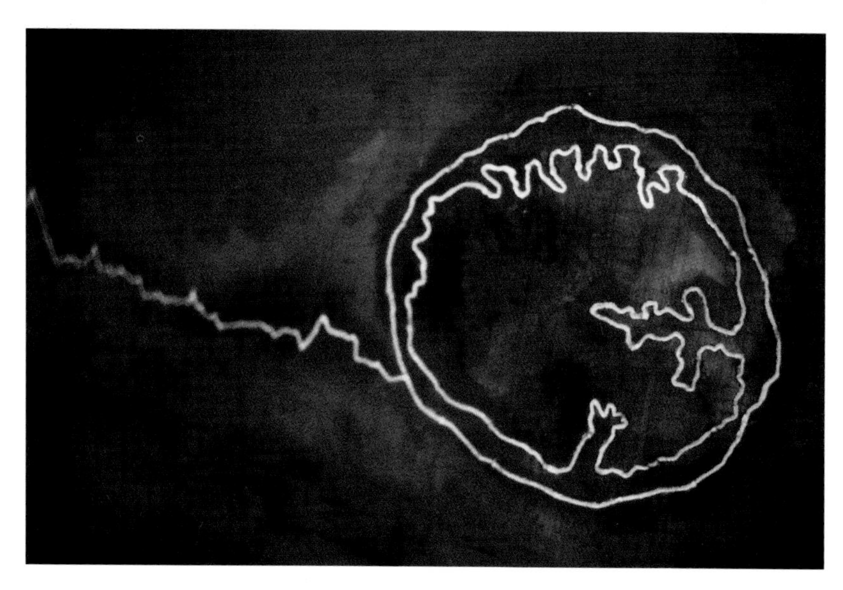

Anna Hill, detail from *Still Like Dust I Rise*, 1996. Luminous brain scan overlaid on a 'human head' formed by natural weathering in yew wood. Yew wood, fibre optics and luminous paint, 35 x 55 cm. Courtesy of the artist. Photo: Mick Williamson.

A vertical section of yew, taken from a wind-fallen tree, has seasoned to reveal the apparition of a human skeleton on which a luminous brain scan is overlaid. Graphic notations suggest scientific measurements. At points on the graph the wood is pierced by tiny red lights, pulsing sequentially, an image reminiscent of ultrasonic brain scans representing cerebral electrical activity in schizophrenic patients.

of image-making, as a fruitful place in which unusual inspiration and even interaction can occur. Pictures and the processes that go into producing them seem, then, to provide a vehicle by which artists and scientists can talk to each other.

A particularly rich location for contemporary image-making, and indeed for knowledge creation, and which has provided just such a neutral cultural space, is information technology. Digital media and computer software have provided the perfect point of contact for technologically orientated artists and artistically sympathetic scientists and technicians.[19]

William Latham's work designing image-generating programmes with an evolutionary quality to them has significantly expanded the manner in which computers can actually generate art forms. Something of the same process can be found in the collaborative work of artist-cum-scientist Richard Brown and scientist Igor Aleksander at Imperial College, on a project in which interactive 3D computer graphics are used to produce 'virtual organisms' which respond to viewers' actions and mimic aspects of the biological workings of the human mind. Another techno-scientific phenomenon set to provide much scope for artistic intervention is that of artificial life (or a-life).[20] A number of other recent ventures, which include the Media Museum in Karlsruhe in Germany (part of the Center for Art and Media Technology – Z.K.M.), the Foundation for Art and Creative Technology in Liverpool, and the InterCommunication Center in Tokyo, as well as the new magazine *ArtByte*, all testify to the likelihood that this area of technological art is set to become a very significant locus of artistic growth in the next millennium. Though there is no space to discuss them here, much of this technological art raises important questions, not least about what actually *is* the art in techno/computer art. Can a computer meaningfully be said to produce art? The most provocative attempts to answer this question with a 'yes' have been pursued by Howard Cohen in San Diego, who has taught computers to 'draw' and now 'paint'. Other interesting works in which computers might more properly be identified as 'collaborators' have been produced, for example, by Jane Prophet (who has made an interactive visual narrative on CD-ROM entitled *Internal Organs of a Cyborg*), and Victoria Vesna, whose project *Bodies inc.* enables visitors to her site to construct and partially control their own ideal bodies.[21]

The implication of projects like these that involve inspired image-makers from both science and art is that significant achievements can result which draw on the best practices of both domains, producing powerful ways of observing or even solving problems. A large number of the creative processes behind collaborative ventures seem to rest on a methodology long felt to be the exclusive domain of the sciences, namely experimentation. But these projects embody a concern to reinvent the 'experiment' in interesting ways, thus, in the

William Latham, *Mutation. Yi Ray Traced on the Plane of Infinity*, 1991. Computer-generated image, using Mutator. © William Latham.

Working at the IBM UK Scientific Centre, Winchester, William Latham has developed a powerful design tool, called Mutator, with which he creates sculptural forms endowed with genetic properties that shape their growth. Mutator is based on Form Grow, a geometrical grammar which uses spirals and fractal recursion to emulate the geometry of natural forms, and provides processes of 'random mutation' and 'natural selection'. Starting from a simple structure, thousands of complex genetic variations can be evolved. The results of such 'Darwinian evolution driven by human aesthetics' are fantastical organisms whose morphologies metamorphose in a sequence of animated images. (Martin Kemp, *Nature*, 26 February 1998.)

words of the Israeli artist Vered Lahav, taking on the mantle of 'visual scientists' doing 'impossible experiments'. It is under the same banner of experimentation that the more forward-thinking galleries, echoing some 1960s ventures of a similar persuasion, have self-consciously performed as laboratories, 'testing grounds for reality' as one commentator has described them.[22]

Images for science, images for art

This chapter has taken us on a tour from the scientific images themselves, through the artistic infatuation with their visual record and finally to the emergence of projects in which artists and scientists use image-making to excite and engage each other. What, then, can we conclude about the role of images in these two domains? Contrary to some contemporary opinion, there is little evidence for the existence of a unified single visual culture or a new 'third way' straddled between sciences and arts. For all sorts of reasons, some of which are discussed elsewhere in this book, the worlds of science and of art remain clearly distinct and firmly separated. However, this chapter has suggested that both the process and the products of image-making in science and art can be enormously enriched by a knowledgeable and sympathetic understanding of both. If the trend toward collaboration is to continue, we may anticipate that a greater exposure to science will enable artists to produce new types of knowledge-filled images and that exposure to art might just help in the evolution of more nuanced scientific image-making and interpretation.

Notes and references

1 Vladimir Nabokov during an interview at Montreux, Switzerland, September 1966. Published in *Wisconsin Studies in Contemporary Literature*, VIII (2), Spring 1967.

2 Martin Kemp, *The Science of Art: Optical themes in Western art from Brunelleschi to Seurat* (New Haven and London, Yale University Press, 1990), pp. 339–41.

3 T.S. Kuhn, 'Comments on the relations of science and art', in *The Essential Tension: Selected studies in scientific tradition and change* (Chicago and London, University of Chicago Press, 1977), pp. 340–51.

4 For the history of photography in science, a good place to start is *Beauty of Another Order: Photography in science*, ed. Ann Thomas (New Haven and London, Yale University Press, in association with National Gallery of Canada, Ottawa, 1997). For an aesthetic tour of SEM pictures see Dee Breger, *Journeys in Microspace: The art of the scanning electron microscope* (New York, Columbia University Press, 1994).

5 See Barbara Maria Stafford, *Artful Science: Enlightenment entertainment and the eclipse of visual education* (Cambridge, MA, and London, MIT Press, 1994), and 'Machines and marvels' in Kemp (1990), see note 2.

6 Stephen Jay Gould, *Bully for Brontosaurus: Reflections in natural history* (New York, Norton, 1991), p. 171.

7 For these and related issues, see Brian S. Baigrie, ed., *Picturing Knowledge: Historical and philosophical problems concerning the use of art in science* (Toronto, Buffalo, London, University of Toronto Press, 1996).

8 See Stanley J. Reiser, 'Technology and the use of the senses in twentieth-century medicine', in *Medicine and the Five Senses*, ed. W.F. Bynum and Roy Porter (Cambridge and New York, Cambridge University Press, 1993). And for the role of pictures in the investigation of the subatomic world, see Peter Galison, *Image and Logic: A material culture of microphysics* (Chicago, Chicago University Press, 1998). Interestingly, Galison's tale ends with victory going to the 'image' rather than 'logic' detectors.

9 Other exhibitions this century include two seminal ones in America: one in 1956 of painting, prints and drawings from the pages of

Scientific American, one surveying images from the cutting edge of science *c.*1965 – both using the same title, *Art in Science.* In 1993 the Paris exhibition, *L'Âme au Corps* took the human body as its subject.

10 The exhibition *Körperwelten* (1997–8) has been shown to crowds in both Germany and Japan.

11 Benedict Rubra worked in an art-science collaboration with Professor Frances Ashcroft of the Department of Physiology, University of Oxford.

12 Maura C. Flannery, 'Images of the cell in twentieth-century art and science', in *Leonardo,* 31 (3), 1998, p. 201.

13 Roberta Smith, 'Galleries are labs of a sort', in 'Arts and Leisure' section of the *New York Times,* 14 February 1999, p. 39.

14 See for example, David Topper, 'Towards an epistemology of scientific illustration', in Baigrie, ed. (1996), p. 235, see note 7.

15 See for example, John Adkins Richardson's chapter on 'Cubism and logic' in his *Modern Art and Scientific Thought* (Urbana, Chicago, London, University of Illinois Press, 1971), and Arthur I. Miller, *Insights of Genius: Imagery and creativity in science and art* (New York, Copernicus, 1996).

16 Anjana Ahuja, 'Science extols Pollock's chaotic art', in *The Times,* 7 November 1997, p. 10.

17 Kemp (1990), p. 1, see note 2. David Topper makes a similar point in 'Natural science and visual art: reflections on the interface', in *Beyond History of Science: Essays in honor of Robert E Schofield,* ed. Elizabeth Barber (Bethlehem, Lehigh University Press, 1990), p. 299.

18 See Martin Kemp's series of 'Art and Science' articles in *Nature,* published as *Structural Intuitions: The 'Nature' book of art and science* (Oxford, Oxford University Press, 2000).

19 György Kepes, *The Language of Vision* (1944), and *The New Landscape in Art and Science* (Paul Theobald & Co, *c.*1956). See also Marga Bijvoet, *Art as Inquiry: Toward new collaborations between art, science and technology* (New York, Washington DC, Baltimore etc., Peter Lang, 1997), pp. 10–22. The Center for Advanced Visual Studies which Kepes founded can be visited at http://cavs.mit.edu. The founding of *Leonardo* (the art and science journal) in 1968 also drew directly on this earlier work.

20 Nell Tenhaaf, 'As art is lifelike: evolution, art, and the readymade', in *Leonardo* 31 (5), 1998, pp. 397–404. Art & Science Collaborations, Inc. have since 1995 held a CyberFair at Cooper Union, NYC, in which various aspects of 'cyberart' are examined and discussed (www.asci.org/cyberart99/index.html).

21 For the impact of modern technology on art, see Jack Burnham in his highly influential work *Beyond Modern Sculpture* (New York, George Braziller, 1968).

22 Smith (1999), p. 39, see note 13, and Ellen Levy, introduction to 'Contemporary art and the genetic code', a special issue of *Art Journal,* ed. Ellen K. Levy with Berta M. Sichel, 55 (1), 1996, p. 7.

5

Us and Them, This and That, Here and There, Now and Then: Collecting, Classifying, Creating

Martin Kemp and Deborah Schultz

> *He who would do good to another, must do it in minute particulars*
> *General good is the plea of the scoundrel, hypocrite and flatterer:*
> *For Art and Science cannot exist but in minutely organized particulars.*
>
> WILLIAM BLAKE, from *Jerusalem.*[1]

Classifying and its limitations

The rudimentary basis of any classificatory system consists of binary oppositions. In our everyday lives, we constantly define and redefine not only others but ourselves according to simple antagonistic pairs of categories. As a ready example, we may cite the way that someone who drives a car oscillates between being a driver and a pedestrian. Let us imagine that the driver passes through a complex city-centre junction in which crawling vehicles and meandering pedestrians are engaged in an elaborate game of barging and bluff. Definitively a 'motorist' – probably a testy one at that – when trying to hustle through the junction, the driver is transformed into an equally impatient 'pedestrian' when traversing the same route on foot. This transformation is typical of the agility and fluidity with which even the same person can repeatedly redefine his or her taxonomic category, with concomitant switches in attitude. It is also possible to occupy plural niches at the same time. Let us imagine that one of the driver's children is in the car, defining the driver as 'father' or 'mother', and, with a high level of probability, as 'husband' or 'wife'. Which of these or a huge number of other potential categorisations comes into play at particular moments depends on the interest of the categoriser and on the context for the act of categorisation.

Time and place are subject to similar fluidities, as a personal example will demonstrate. Let us imagine that one of the authors visits St Andrews, where they both used to work, and that he or she bumps by chance into a former colleague who asks, 'where are you now?' The answer is, 'I've been in Oxford for three years now.' In this exchange, Oxford thus becomes 'here' or 'there' in that general sense of 'now'. But if he or she telephones someone else in Oxford, and asks if that person would like to come over to see him or her, the obvious question is, 'where are you now?'. The previous reply, 'Oxford', is too non-specific, both with respect to location and time. The taxonomic category of 'here' needs to be the new office in George Street, the old office in Beaumont Street or some other specific location. 'Now' means at this moment and for the immediate future, the immediate future being defined in terms of how long it will take the other person to reach the author's present location.

What these examples are designed to show is that our fundamental habits of binary differentiation are incredibly subtle and context-determined even in simple daily use. The categories become functional through a remarkable and generally unstated 'contract' of shared terms of reference between two or more individuals. It is apparent that taxonomies in art – even the big oppositions of 'Art' and '*not* Art', or 'Art' and 'Science' – are of the protean nature of us and them, here and there, now and then. But what of our need, also deeply functional, to form more watertight categories which resist the fluidities of the binary oppositions we have illustrated so far? The overt desire for taxonomies which exhibit quasi-mathematical characteristics, such that they can be tabulated or expressed graphically in terms of branching systems, is especially associated with Western science, but it is present in any society which exhibits complex behaviour in relation both to the external world and to the artificial structures contrived by humans. In such contexts, many types of made or 'found' objects acquire precisely defined functions – ranging from utilitarian to magical – so specific to those objects that they require clear definition and exclusive identity in relation to otherwise comparable items. To take a case that has become familiar through Gombrich's 'cupology', a cup is made under the tacit classification of 'cup', which itself may be seen as belonging to various categories of cup according to the interests of the classifier, such as 'cups with and without handles', or 'earthenware and porcelain' and so on.[2] Even here, we will encounter a complex series of possibilities according to the conscious and unconscious motives of the producer of the items and of their subsequent categorisers.

The kind of mathematical precision in taxonomy that aspires to eliminate such fluidities to an extreme degree became the prerogative of Western science, above all in the phases when it was crystallising in its modern intellectual and institutional guises. Historically it is exemplified by the progressive triumph of Linnaeus's binomial system in the second half of the eighteenth century over the kind of human-orientated 'natural order' favoured by the Comte de Buffon, whose *Histoire naturelle … Description du Cabinet du Roy*, issued in no fewer than 44 volumes from 1746 onwards, remains an unrivalled monument of orderly naturalistic description. Linnaeus's classificatory naming of plants by genus and species according to tabulated accounts of their sexual organs – progressively detached from the Romantic context in which it particularly flourished – possessed a 'harder' feel that appealed to later generations of professional scientists. Other needs may well generate alternative classifications that are no less systematic and functional. A nice instance is the 'bird book' that serviced the needs of many aspiring ornithologists in the 1950s, the *Pocket Guide to British Birds* by Fitter and Richardson, which was designed specifically for field identification. Using a key arranged in four main divisions – Plumage, Structural Features, Behaviour Features, and Habitats (with a Migration Table) – it was possible through processes of elimination to arrive at the one and only species that would 'fit' with the observations.[3] This field key was not intended to supplant the biologists' classification of the Linnaean kind, but was designed to provide a practical tool for the naturalist which has much in common with de Buffon's more relaxed process of organisation.

In recent years, the challenge to our assumption that any such classificatory orders are 'given' by the nature of things has been led by Foucault – in effect extending de Buffon's argument that any elaborately systematic taxonomy is a mental contrivance ordained by the predispositions of the classifier. The *locus classicus* is the famous passage in Foucault's 1966 publication, *Les Mots et les Choses*, in which he quotes the much cited passage from Borges about a 'certain Chinese encyclopaedia' that divides animals into categories remote from our taxonomies – including 'belonging to the Emperor', 'drawn with a very fine camelhair brush', and 'fabulous'.[4] However, Foucault's delighted and skilful use of the exotic text to indicate an incommensurability of taxonomic interests and values in different cultures cannot be accepted at face value. As Martin Kemp has argued elsewhere:

> The cited presence of the category 'stray dogs', for example, suggests that there was something recognisable as 'a dog' rather than as 'a cat', and that strays were recognised then as now. Even the apparently improbable 'drawn with a very fine camelhair brush' is not so very remote from what we do when we classify a Dürer watercolour of a bird's wing as 'a drawing' and file it away under 'drawings of animals' – or 'of birds' or 'of parts of animals', depending upon how many items we have and with how fine a sieve we wish to strain the material. As with our Dürer classification, we have to ask by whom something is being classified and for what purpose in what functional context. [5]

Moreover, the problems do not stop here. Attempting to trace the original Chinese source takes us on a frustrating and ultimately futile paper-chase – like the search for the chimerical data about the Eskimo's supposedly multiple words for snow, which has become a repeatedly cited mantra of those who prefer to subjugate the visual experience to the verbal.[6] Foucault's immediate source is Borges's essay, 'El Idioma Analítico de John Wilkins', in which he discusses the arbitrariness of classifications through a study of the seventeenth-century philosopher, who had been omitted from the 14th edition of the *Encyclopaedia Britannica*.[7] Borges writes that the 'ambiguities, redundancies and deficiencies' of universal schemes of classification 'are reminiscent of those which Dr Franz Kuhn attributes to a certain Chinese encyclopaedia entitled the *Celestial Emporium of Benevolent Knowledge*'. Borges's reference to Kuhn (1889–1961), a prolific translator from Chinese, has proved stubbornly elusive, and is probably one of the great author's typical teases. In any event, it is clear that the taxonomy 'cited' by Borges bears little resemblance to the major Chinese systems for classifying the natural world, which share much in common with our later orders.[8] It is increasingly looking as if the Foucauldian insistence on the cultural arbitrariness of all taxonomies stands in need of substantial qualification.

Although the conceptual and perceptual grounds on which the profusion of animals and plants has been placed into orderly schemes have changed substantially over time and across cultures – not least within evolutionary and non-evolutionary frameworks – they can all be seen to share a certain measure of structural similarity with respect to how organisms occupy definable niches and stand in relationship to humankind. It is becoming increasingly clear in ethnobiology that this robustness can be discerned across a wide range of modern and traditional societies.[9] Our present system of biological taxonomy, which centres on the metaphor of the evolutionary 'tree' is a non-arbitrary response to the types of surviving and extinct animals.[10] At the same time the system can be seen as possible in that precise form only within the conceptual and institutional frameworks of a biological science in which evolution is the dominant paradigm, particularly given an implicit value-system in which *homo sapiens* is placed at the top of the evolutionary tree.[11] Hierarchies tend inexorably to accompany taxonomies, as surely as our metaphorical ladders have upper and lower rungs. As soon as there is a human dimension in any scheme, 'this and that' take on the colours of 'us and them' and 'higher' and 'lower'.

Subversive strategies in Broodthaers's *Section des Figures*

It is easy to see why such classificatory systems have come to be regarded simultaneously as expressive of particular social systems and as hostile to creative thought. Precise tabulation and binomial nomenclature can, particularly in a pedagogic context, all too easily become a form of intellectual stamp-collecting – a symptom of imaginative constipation. It is not surprising that many of the significant moves in twentieth-century art have relied upon self-proclaimed subversion of standard taxonomic categories within and outside the category of 'Art'. The most noted of these moves in the twentieth century was the introduction of 'ready-mades' by Marcel Duchamp. When he signed the urinal 'R MUTT', he was both denying the validity of the category of 'Art' through which we bring certain expectations to our viewing of certain types of objects, and, as has become clear in retrospect, was radically expanding the category of what we can choose to call 'Art'.

Duchamp's challenge to a binary system of 'Art' and 'Non-Art' has echoed repeatedly throughout the twentieth century, as successive iconoclasts have breached the boundaries of art or assaulted the very notion of art itself. The most cogent of the many barbs directed at the artificiality of museological categories was the series of manifestations from the *Musée d'Art Moderne, Département des Aigles*, devised by the Belgian poet and artist Marcel Broodthaers. In 1968, at a time of political unrest across Europe, Broodthaers inaugurated his *Musée*, from which he subsequently conducted 'official' correspondence like a curator.[12] The museum underwent a series of transformations, and in 1971 the *Section Financière* announced its impending sale due to bankruptcy. The most overtly museum-like of its metamorphoses was staged at the Städtische Kunsthalle in Düsseldorf from May to July 1972, where the *Section des Figures* comprised 282 objects each bearing the image of an eagle. It served as a kind of latter-day *Wunderkammer* or cabinet of curiosities for both natural and artificial objects. Whilst some were indisputably 'works of art' (for example, Magritte's *Fontaine de jouvence* of 1957, which depicted an eagle of stone, or a specially commissioned painting by Gerhard Richter), others ranged from jewellery and ceramic pots to logos on wine labels, or on typewriters. There was also 'the real thing' in the form of stuffed birds. All the objects had been

Marcel Broodthaers, View of the *Musée d'Art Moderne, Département des Aigles, Section des Figures*, 1972. Courtesy of the Marcel Broodthaers Estate. © DACS 1999.

borrowed from either museums or private sources and were returned at the end of the exhibition.

Broodthaers broke with conventional classification regarding both the range of exhibits and the nature of his own activity. A poet, who had very publicly launched himself as an artist in 1964, he now took on the role of curator and, instead of making works to be exhibited, borrowed pre-existing objects. His own 'work' was the exhibition itself rather than the contents exhibited. To emphasise the various origins of the items he listed a number of the museums on the front and back covers of the two volumes of the catalogue. Inside he provided details about all the objects and their ownership, presumptuously listing the *Département des Aigles* alongside established museums. On the cover of the first volume was a photograph of three eagles' eggs with their original museum labels. It was an image which neatly encapsulated the main underlying themes of Broodthaers's work at this time – the shell, the museum, classifying, the word-image relationship, the symbol of the eagle, and repetition.

By bringing together such apparently heterogeneous objects under his chosen heading, Broodthaers was commenting upon the ways in which museums departmentalise their stock and exclude non-collectable items. In the 1974 interview/text, 'Dix milles francs de récompense' ('Ten Thousand Francs Reward'), the 'interviewer' referred to the much quoted line on chance juxtaposition from the writings of the Comte de Lautréamont (the poet Isidore Ducasse, who lived from 1846 to 1870): 'This sort of claim to embrace artistic forms as far distant from one another as an object can be from a traditional painting – doesn't it remind you of the encounter of a sewing machine and an umbrella on an operating table?'[13]

Broodthaers's reply distanced himself from this Surrealist approach. Instead, he indicated that his choice was based less upon juxtaposition and more upon a studied disruption of conventional forms of classification:

> A comb, a traditional painting, a sewing machine, an umbrella, a table may find a place in the museum in different sections, depending upon their classification. We see sculpture in a separate space, painting in another, ceramics and porcelains ..., stuffed animals ... Each space is in turn compartmentalised, perhaps intended to be a section – snakes, insects, fish, birds – susceptible to being divided into departments – parrots, gulls, eagles.[14]

The Düsseldorf exhibition, then, seems to have little to do with Surrealist juxtaposition and more in common with Foucault's *Les Mots et les Choses*, the English title of which, *The Order of Things*, is thought to be Foucault's original. The assemblage of 'eaglery' in the *Section des Figures* signals a questioning of the established order of things, and is indicative of Broodthaers's practice. Whatever the historical status of Foucault's 'Chinese encyclopaedia', the unconventional taxonomy may be taken as signalling 'another system of thought' – a system that indicated 'the limitation of our own, the stark impossibility of thinking *that*'.[15] Not accepting the constraints of 'impossibility', Broodthaers was precisely not afraid to think *that*.

Within his overarching challenge to the 'scientific' taxonomies of museums, Broodthaers also wished to undermine contemporary definitions of the 'art object'. Beside many of the objects he placed small plastic labels on which was printed a number and 'This is not a work of art' in English, French or German. As he explained in the catalogue: 'Whether a urinal signed "R. Mutt" (1917) or an *objet trouvé*, any object can be elevated to the status of art. The artist defines the object in such a way that its future can lie only in the museum. Since Duchamp, the artist is author of a definition.'[16]

However, Broodthaers regretted that 'Duchamp's initiative' had lost its 'destabilising' impact, such that 'having become a mere shadow of itself – it dominates an entire area of contemporary art, supported by collectors and dealers'.[17] Broodthaers would take Duchamp further by presenting such definitions as labels which can just as easily be removed as applied, during or after the assembly of his temporary collection.

The repetitive image of the eagle underlines the relationship between the material and the poetic in Broodthaers's work. Originating from a 1957 untitled poem, the name *Département des Aigles* was 'a literary memory'.[18] However, in the *Section des Figures* Broodthaers constantly shifts between a use of metaphor and a materialist approach. The second volume of the catalogue contained an essay by the anthropologist Michael Oppitz, whom Broodthaers had invited to contribute in order to add a more 'theoretical' dimension to the

Marcel Broodthaers, Vitrine with label 'This is not a work of art' from the *Musée d'Art Moderne*, 1972. Courtesy of the Marcel Broodthaers Estate. © DACS 1999.

discussion. In his essay, Oppitz described the 'heavy symbolic burden' of the sign of the eagle, its 'emblematic, mythological significance. It variously connotes strength, virility, rigour, freedom, authority, aspiration towards the absolute'.[19] Broodthaers himself described the eagle as, 'grandeur, authority, strength. Divine spirit. Spirit of conquest. Imperialism'.[20]

Notwithstanding these conventional and extraordinarily widespread connotations, Oppitz then signalled the desensitising effects of Broodthaers's classificatory re-ordering: 'By identifying the symbolic presence in every conceivable eagle, Broodthaers engages in an incessant defusing of the eagle's mythic power.'[21] The cool, pseudo-scientific nature of display cases and vitrines, together with the labels, emphasised the corporeal existence of the exhibits and dispelled their mythical, immaterial qualities. In contrast with Duchamp's urinal and Magritte's pipe – everyday objects which, according to Oppitz, had become 'cherished fetishes'[22] – Broodthaers reduced all the exhibits to the level of 'object'. Broodthaers's alternative form of classification would, therefore, have a profound consequence upon the objects exhibited, altering the way in which they would be received and thereby emphasising the fundamental importance of contexts and relationships. Broodthaers was in effect arriving at a new kind of 'natural order', based on the relationship between things and us, like a latter-day de Buffon, rather than accepting the established science of deeply established taxonomies of material objects within our ordered systems of knowledge.

Broodthaers's poetry permeated his work in the visual arts. His approach combined poetic metaphor with the rationality of Conceptual Art. The use of fictions characterised Broodthaers's method of questioning and enabled him quietly to subvert existing structures. He set up his fictive museum in order to challenge 'real' museums for, 'with the help of a fiction like my museum it is possible to grasp reality as well as that which reality conceals'.[23] As will be seen with the other artists we will be examining in this context, radical acts of de-classification simultaneously become acts of re-ordering that make the represented items 'look' notably different, both in themselves and collectively.

On relics and value: Hiller at the Freud Museum

A more personalised act of re-collecting or recollecting – an interesting and significant word – occurred more recently when the American artist Susan Hiller, who is resident in London, undertook a sustained dialogue with the collection of Sigmund Freud, on display until 1996 in the house in Hampstead that the pioneer of psychoanalysis occupied for fifteen months of his final exile in England.[24] Freud's collection, in which each item is documented with curatorial care, is conventional in the sense that its combination of antiquities (predominantly Egyptian, Classical and African) could have belonged to any moderately well-to-do European collector of his generation. What Susan Hiller accomplished – in a house now re-labelled 'Museum' – was to generate an alternative collection which is genuinely Freudian in spirit, incorporating elusive traces of memory, allusion and personal association in a free flow of implicit narratives which the viewer can write or rewrite on his or her own terms. In so doing, she highlights the fact that every collection of objects embodies both obvious and hidden narratives of potentially multiple kinds – both for its assemblers and its successive viewers. Trained as an anthropologist and educated in archaeological procedures of gathering and storage, Hiller uses conventional museum techniques to show how collections of labelled objects and images service the archaeology of memory in ways beyond the ostensible rationale for their assembly.

Hiller's 25 'scientific' containers – uniform brown cardboard boxes of archival standard – were displayed open in a vitrine. Each box was labelled, often in a language other than English and accompanied by translations. For Hiller the working process she adopted was 'dreamlike' in the sense that there was both a manifest and a hidden relationship between the component parts of the boxes which 'present the viewer with a word (each is titled), a thing or object, and an image or text or chart, a representation. And the three aspects hang together (or not) in some kind of very close relationship which might be metaphoric or metonymic or whatever.'[25]

Characteristically, Hiller suggests possible associations, but eschews any absolutely defined relationships. She allows links to remain unclear, questions to hang unanswered. There is a constant mediation between the 'dreamlike' and the scientific in both the choice and display of exhibits. One box in particular, *Seance/seminar*, combines these often opposing qualities. It contains a photo-

Susan Hiller, from *At the Freud Museum*, 1991–6. Courtesy of the artist. In the collection of the Tate Gallery, London.

copied engraving from the *Ars Magna Lucis et Umbrae* (1671) by the Jesuit encyclopaedist and purveyor of arcane mysteries Athanasius Kircher. The Kircher engraving shows the prototype of a magic lantern, one of the kinds of device that he used to inculcate awe in 'natural magic', together with a miniature LCD monitor showing *Bright Shadow*, a video by Hiller in which shadowy images were formed from light passing through a shuttered window. As Hiller wrote: ' I've been collecting earth, water and in this box, shadows.'[26] The intangible also becomes collectable, and is preserved on video.

In her 1996 retrospective at the Tate Gallery, Liverpool, Hiller projected the video onto the floor of a dark, enclosed area, thereby drawing direct parallels with the room-sized camera obscuras illustrated in Kircher's volume.[27] Hiller and Kircher's walk-in spaces serve as chambers of visual magic in which projected images undergo mysterious transformations. The title, *Seance/seminar*, emphasises the 'dreamlike'/scientific approach, as well as the unstable relationship between language and meaning: 'A seance, which is supposedly irrational and untheorized, means in French "a seminar", which we think of as respectable and academic and logical.'[28]

Hiller's box on the uncanny, *Heimlich/homely*, ideally suited to the Freudian context, highlights the impossibility of a wholly accurate translation of a loaded term from one language to another, and the blurring of contradictions between supposedly opposing terms. The box contains an old record of 'Look Homeward Angel' sung by Johnnie Ray, the title of which is derived from a novel by Thomas Wolfe, in which the angel is a stone image in a graveyard. Home is, by implication, death. As Freud explained in his essay on the uncanny, the meaning of *unheimlich* is actually very close to that of *heimlich*. Therefore, *heimlich* functions as a developing concept rather than an oppositional term: 'homely' carrying connotations of cosy, private, secret, furtive and incestuous. The straight English translation of *unheimlich* as 'uncanny' is clearly inadequate, and surrenders a whole range of inherent meanings. Hiller overcomes the apparent strictures of oppositional relations and translations by eliding them visually within her work. As she explained: 'I confess a long standing wish to blend art and science, poetry and analysis.'[29]

The format of the display used in the Freud Museum reflects Hiller's background as an anthropologist when she would gather materials as objective data. However, she manipulates the scientific, museological display format in a very particular way. She does not claim objectivity, which she considers 'a fantasy our culture is heavily invested in', but, on her own behalf, she needs to gain the trust of the viewer.[30] This is apparent in a box such as *Eaux-de-Vie/spirits* in which she encloses a text by Robert Graves with spring water she has collected in antique bottles from the Lethe and Mnemosyne (sources in the gorge of Herkyna which respectively obliterate and imprint memory). The treasured samples have then been corked, sealed and tagged. The viewer is required to believe the label and the tags, as there is no *visual* evidence to show that the bottles do not contain ordinary tap water. The water looks like any other; its authentication is certified by the context and the textual information – the words, the bottles, the labels, the cases – which tell us that the water must hold special significance. The level of trust is like that demanded by the relic of a saint – say a finger, indistinguishable in itself from any other severed digit.

The text by Graves refers to the Greek myth which decrees that we should avoid drinking the forgetful water of the Lethe in favour of the memorable Mnemosyne, in our quest to become immortal and indestructible. Memory is central to Hiller's collection and her selection of exhibits: 'My starting points were artless, worthless artefacts and materials – rubbish, discards, fragments, trivia and reproductions – which seemed to carry an aura of memory and to hint at meaning something, something that made me want to work with them and on them.'[31]

The Freud Museum project relies upon an excavatory process which may not always unearth hidden meaning but which ensures that the materials are accredited with implicit value. Reproductions, even cheap photocopies, are used rather than original images. Objects and materials are selected which may be unusual but which are not unique. The mundane is rendered special in Hiller's 'archaeology of the domestic'. She presents examples of things which are around us but which we often overlook because they do not rate highly enough in the hierarchy of values within our culture. She does not materially alter the exhibits but shows them as they are, for any intervention on her part would not add to their value in the context she creates. She forces us to look again and to be aware of the cultural codes within which we observe the world around us. Hiller uses art in order to do this for: 'I think what art does is to reveal hidden, undisclosed, unarticulated codes within a culture.'[32]

The various works exhibited in the Freud Museum are part of an ongoing project to which she continues to add suitable items. Operating in a related manner to *Eaux-de-Vie/spirits* is *Nama-ma/mother* which contains a photocopied diagram of Australian cave paintings from Uluru or Ayers Rock, and some samples of natural pigments collected by Hiller near Papunya, in the Australian desert, and stored in small perspex cosmetic containers. Here Hiller examines the indecipherable nature of unknown languages – both the aboriginal word for mother and the diagrammatic drawings of the cave paintings. She links the pigments, classified as 'native earths' with the notion of mother earth. Again, the faith of the viewer is essential if the contents of the containers are to be credited as significant and not just seen as any old dust.

Vehicles of display familiar from science serve magical ends in Hiller's open-ended collection. Whereas Broodthaers deconstructs the mythical packaging of the eagle, Hiller assembles a whole range of devices to construct value and mystery around materials and objects which are often visually insignificant in themselves. The process is akin to that of the decorative mounting of the objective lens from Galileo's telescope, presented to Grand Duke Ferdinand II

Susan Hiller, *Nama-ma/mother*, 1991–6. Detail from *At the Freud Museum*. Courtesy of the artist. In the collection of the Tate Gallery, London.

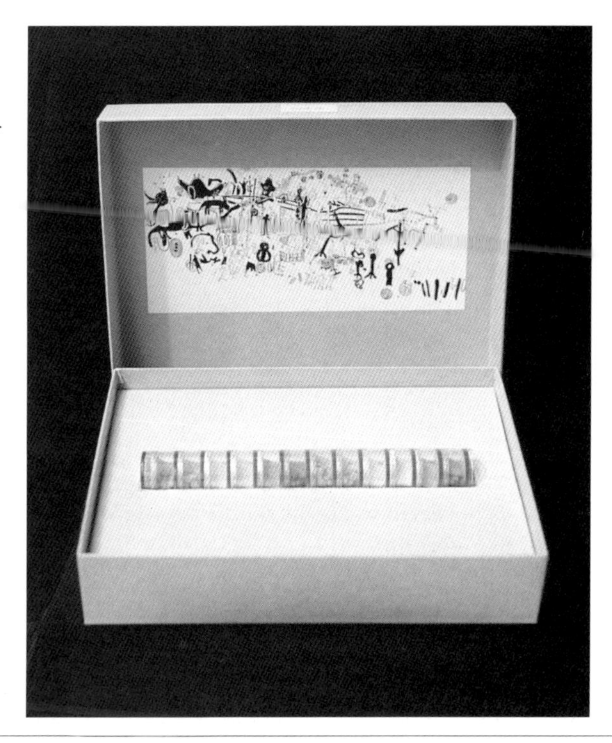

of Tuscany and piously preserved in Florence within an elaborate and inscribed ivory frame made by Vittorio Croster – though materially and visually the lens is just an old piece of cracked glass.[33]

de vries: recording the unnamed

Hiller's set of pigments from 'mother' earth parallels the activity of another artist-gatherer, herman de vries, assembler of things of a kind, avoider of capitalisation, philosopher and poet. de vries's strategy is, however, less concerned with metaphor and memory than Hiller's, and assumes methods more akin to those in the herbaria and arboreta in which he has often worked.[34] On the face of it, the systematically arrayed collections of shells, flowers, leaves, branches, herbs and varieties of earth which the Dutch artist has been assembling since 1970 recall the reference collections in natural history museums and botanic gardens, stored in orderly rows in trays and drawers. However, his subtle and often intuitive re-groupings of naturally occurring things quietly but decisively subvert the stock categories of scientific order and resist the taxonomic processes of naming. His montages of multitudinous samples of earths characteristically alert us to an astonishing variety of types which we can differentiate visually but which far outstrip our restricted vocabulary of colour terms and other descriptors.

By 1992, he had assembled a collection of more than 2,400 samples of earth, including no fewer than 300 from Groningen in Holland and 350 from the volcanic island of Gomera in the Canaries. de vries lays out the samples in neat grids, according to criteria of similarity and difference, but not within rigidly predetermined categories. The nuanced visual differences in the world-wide manifestations of the ancient 'element' of earth speak by implication of the manifold variations in the lives of the organisms that the element supports, not least of the wonderful attuning of plant physiologies to minute variations in the composition of soil.

de vries's art involves the subtle contemplation of varieties, often expressed within very narrow compasses of difference. In its quiet way, his art is as subversive as Broodthaers's. When he exhibits collections of herbs, rich in odour and taste, he plays to sensory perceptions even more resistant to exact naming than objects of sight. The hundreds of herbal substances from exotic and homely locations exhibited under the title *natural relations* in 1989 make manifest an enduring aspect of human relationships with plants in a way that comprehensively crosses geography and time. When he records the sounds of the world he alerts us to a natural music which does not lend itself to conventional notation. When he writes without capital letters, as he has done for 20 years, he intends to signal his antagonism to the hierarchies which traditional naming and taxonomies insidiously reflect. Looking at nature, we can see that:

> every part has its own function, so why should a tree be more important than a diatom?... language is 'you and me', 'we and them', 'here and there',

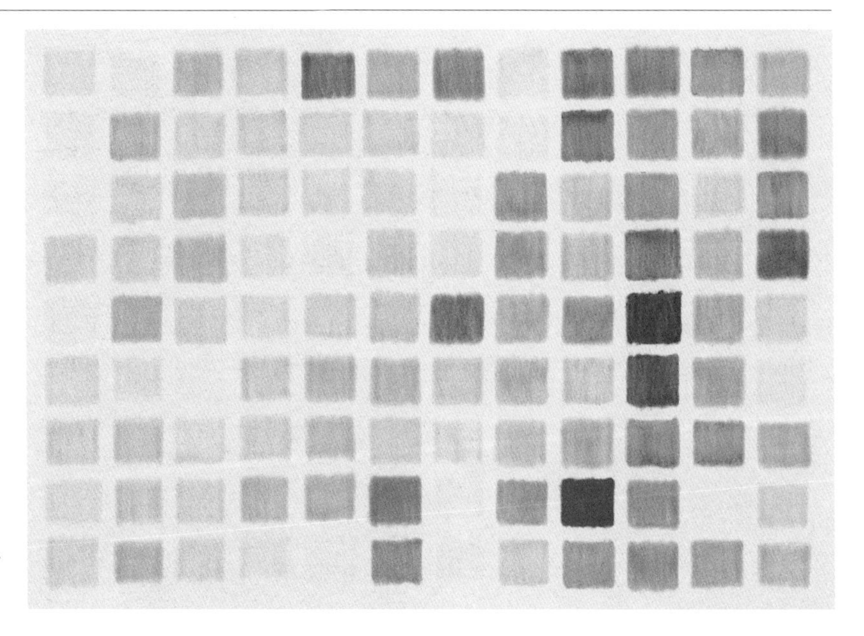

herman de vries, *from earth; nepal and india*, 1992. Earth on paper, 100 x 140 cm. Courtesy of the artist. Photo: Bruno Schneyer.

herman de vries, *natural relations*, 1982–9. Dimensions variable. Courtesy of the artist.

> whereas in effect, it is all part of the same, it's one. language ... gives us a grip on reality and great social power, but we also pay for it in the loss of unity.[35]

de vries's book of nature contains no rigidly prescribed lists of named flora and fauna in set ranks and rows. He invites us to re-see within refreshed frames of perceptual reference, so that we can sense affinities across the hermetically sealed boundaries characteristic of conventional classifications. His is a science that opens rather than shuts categories, allowing relations between discrete items to express holism rather than separation and opposition.

Fleming's compilations, Parker's transformations

The artist as collector-exhibitor-classifier-curator-designer has become an increasingly common feature in the making of installations – the type of exhibit that has become one of the most characteristic art forms of the last third of the century. The range of approaches in the assembling and re-working of collections has been remarkably diverse – as exemplified by the different conceptual and visual orientations of Broodthaers, Hiller and de vries. The extremes can be surprisingly wide apart. At one end of the spectrum is the decisive – even explosive – physical intervention in the material form of collected items as practised by Cornelia Parker in some of her larger works. At the other is the surreptitious insertion of surrogate or alien 'collectables' amongst cased displays of 'real' museum objects, such as Richard Wentworth's scrupulously labelled collections of common-or-garden detritus from the basement of the British Museum in 1997, assembled in standard museum wall-cabinets and reticently secreted in the august setting of the Egyptian Galleries.[36]

A particularly effective instance of the latter approach is the project *Atomism and Animism*, which has resulted in 1999 from Martha Fleming's residency in the Science Museum, London, and has involved decorous insertions of 'foreign' items in thematic cases and suggestive compilations of diverse objects from different departments in the Museum and even from non-scientific collections housed elsewhere.[37] For example, in a handsome museum display case which she has dedicated to a compilation of conical items, the viewer is presented with objects of such diverse function as shuttlecocks from the game of badminton, a megaphone, filter-funnels, a giant hourglass and mathematical

Martha Fleming, *Spheres* and *Anemic Cinema*, 1999. Museum display case containing circular and spherical objects, in the exhibition *Atomism and Animism* (1999), the culmination of the artist's residency in the Science Museum, London. Courtesy of the artist and National Museum of Photography, Film and Television/SSPL.

models that demonstrate the beguiling geometry of conical bodies. In a case where the top shelf is entitled *Spheres* and the lower shelf *Anemic Cinema* – taken from Duchamp's name for the apparatus of spinning disks which he used to create optical effects – a cinematographic disk from Muybridge's zoopraxiscope is accompanied by all manner of circular and spherical objects, many on well-turned stands, while another compilation invites us to reflect on the relationship between the ellipse and the circle, and the ellipsoid and the sphere. The status of the ellipse in relation to the circle had been irreversibly rewritten by Kepler's demonstration that the orbits of the planets are elliptical rather than conforming to a perfect circle.

Fleming notes that this relationship was the subject of quoted meditations by both Einstein and Wittgenstein. Einstein posed a specific challenge when answering critics of his theory of relativity:

> Let the surface of S1 prove to be a sphere, and that of S2 an ellipsoid of revolution. Thereupon we put the question – what is the reason for this difference in the two bodies? No answer can be admitted as epistemologically satisfactory, unless the reason given is an *observable fact of experience* [but] ... the physical system consisting of S1 and S2 reveals in itself no imaginable cause to which the differing behaviour of S1 and S2 can be referred. [Therefore] the mechanical behaviour of S1 and S2 is partly conditioned in quite essential respects, by distant masses which we have not included in the system under consideration.

Wittgenstein, for his part is characteristically elliptical in his mode of argument:

> An historical explanation, an explanation as a hypothesis of development, is only one kind of summary of data – of their synopsis. We can equally well see the data in their relations to one another and make a summary of them in a general picture without putting it in the form of a hypothesis regarding the development over time ... This 'all round' type of presentation makes possible that understanding which consists just in the fact that we 'see the connections'. Hence the importance of finding *intermediate* links. But in our case an hypothetical link is not meant to do anything except to draw attention to the similarity, the connection between the *facts*. As one might illustrate the relation of a circle to an ellipse by gradually transforming an ellipse into a circle: *but not in order to assert that a given ellipse, historically, came from a circle* (hypothesis of development) but only sharpen our eye for a formal connection.[38]

While Einstein formulates hypothesis of difference in terms of physical action beyond the system, the philosopher posits conceptual links which connect the different configurations together, without any physical process of transformation.

Fleming invites us to think beyond obvious questions of the form, function and name of the individual, unitary objects which pertain to cones and conic sections, and to extend our imaginations laterally into unexpected similarities and variations in the mental and physical life of things. The conjunction of a shuttlecock and filter-funnel, for example, seems at first sight to be whimsical, like the encounter between Lautréamont's sewing machine and umbrella, but *if we think about it*, there may be something suggestive and even profound about the morphology of conical items designed to operate dynamically with fluids. There may be connections in both the Einsteinian and Wittgensteinian senses. The processes of insight which the viewer is invited to activate share a deep affinity with the kind of lateral leaps of imagination that lie behind those scientific discoveries in which unexpected conjunctions or fresh analogies stimulate the formulation of new hypotheses.

At the other pole, Parker's *Cold Dark Matter* serves to illustrate how collections of items can be handled and presented when the items themselves are subject to transformative intervention.[39] *Cold Dark Matter* bears witness to the encyclopaedic reach of the associations that she embeds in her collected objects, a reach that extends from childish play to the cosmos of science – and, in this sense at least, she inhabits a comparable territory to our other artists. To

Cornelia Parker, *Cold Dark Matter*, 1991. Courtesy of the artist. In the collection of the Tate Gallery, London. Photo: Hugo Glendinning, courtesy of Chisenhale Gallery, London.

begin with, an inoffensive garden shed, illuminated from within and stuffed with a collection of acquired clutter, was located peacefully in an empty gallery. Then, courtesy of the Army, the shed was blown asunder in a field, its contents violently scattered and re-configured – a blasted plastic dinosaur, a ruptured boot, a torn hot-water bottle, a euphonium wrenched out of tune, and so on. Laboriously re-collected, the debris was reassembled in the gallery as a galaxy of fragments hanging on wires. Illuminated by a bright inner light, the strange constellations cast eerie shadows on the confining walls, roof and floor.

Parker is not *illustrating* the scientific concept which supplies her title, *Cold Dark Matter*. Rather she is exploiting the metaphorical poetry inherent in many of science's apparently cool and calculated acts of naming and description. The moment of catastrophic expansion is frozen in a state of uncertainty, suspended between total disintegration and potential reconstitution. Are we witnessing the Big Bang or the gravitational implosion of a black hole? *Cold Dark Matter* orbits around binary categories of gathering and dispersal, expansion and contraction, explosion and implosion, inhaling and exhaling, falling and suspension, transformation and stasis, peace and violence, playful and sinister – existing in an indeterminate state and available for alternative perceptions by the observer.

The obvious moral to be gleaned from the examples of challenges to taxonomic norms by the artists we have encountered is that there is a binary opposition between set classificatory categories and the kind of imaginative leaps that creativity demands. Indeed, this view has become something of a conventional norm – correlated with the idea that the artist's job is to rescind intellectual, aesthetic and institutional boundaries erected by the categories to which conventional lives and attitudes are subject. However, the real moral seems to be different from this interpretation – and even opposed to it. What all our de-classifiers are doing is to re-classify so that we can look afresh. They are not setting up fixed, absolutely predetermined taxonomies, to be sure, but neither are they providing us with conjunctions of an entirely arbitrary nature. They show how acts of re-ordering can exercise profound and creative effects on how things are seen, and can themselves result in new discoveries.

There is no more striking exemplar of this moral than Dmitry Mendeleev's Periodic Table of the elements, inaugurated in the late 1860s, which resulted in the reconfiguring of a whole science.[40] The creative potency of Mendeleev's table was not just that it functioned as a tool for arranging elements in accordance with their properties but that the gaps in the sequences predicted 'the discovery of yet *unknown* elements', as first happened with gallium. For Mendeleev, as a philosopher-chemist, who wrote on such things as the 'Unity of Matter', the success of his table triumphantly proclaimed the value of what we would call theoretical modelling in the face of the narrow empiricism of known facts. He proclaimed that 'sound generalisation – together with the relics of those which have proved to be untenable – promote scientific productivity, and ensure the luxurious growth of science under the influence of rays emanating from the centres of scientific energy'.

In a sense, all our artists are Mendeleevs, not only inventing new orders that serve to rearrange our perceptions but also inviting us to discover as yet unknown elements within the frameworks they have set up. The enemy of creativity in the arts and sciences is not classification. The enemy is the unthinking acceptance that any given taxonomy provides us with access to everything that we can possibly see and do. A fresh set of taxonomic conjunctions unleashes the potential for liberated thinking and looking, in art no less than in science.

Notes and references

1 William Blake, *Jerusalem*, 1815, chapter 3, plate 55.

2 For the cod-discipline of 'cupology', see E.H. Gombrich, 'Approaches to the History of Art: Three Points for Discussion', in *Topics of Our Time* (London, Phaidon, 1991), esp. pp. 66–7.

3 R. Fitter and R. Richardson, *Pocket Guide to British Birds* (London, Collins, 1952).

4 The source for English-speaking readers is M. Foucault, *The Order of Things: An archaeology of the human sciences* (London and New York, Tavistock, 1970), p. xv.

5 M. Kemp, 'The ambiguous object: the perception of artefacts within changing taxonomies', in *The World, the Image and Aesthetic Experience: Interdisciplinary perspectives on perception and understanding*, ed. C. Murath and S. Price (Bradford, University of Bradford, 1996), pp. 193–222. See also M. Kemp and Marina Wallace, 'What's it Called and What's it For?', in *Interfaces*, XIV, 1998, pp. 9–25.

6 For this cautionary tale, see G. Pullam, 'The great Eskimo vocabulary hoax', in *Natural Language and Linguistic Theory*, VII, 1989, pp. 275–81, drawing upon L. Martin, '"Eskimo words for snow": a case study in the genesis and decay of an anthropological example', in *American Anthropologist*, LXXXVIII, pp. 418–22.

7 J.L. Borges, 'El Idioma Analítico de John Wilkins', in *Obras completas* (Buenos Aires, Emecé Editores, 1974), p. 708; and E.R. Monegal, ed., *Ficcionario* (Mexico, Fondo de Cultura Económica, 1985), pp. 184–5. For the context in Borges's thought, see D.T. Jaén, *Borges' Esoteric Library* (New York and London, Lanham, 1992), pp. 51–4.

8 For a discussion of Chinese classifications of nature in the encyclopaedias, see J. Needham's *Science and Civilisation in China* (Cambridge, Cambridge University Press, 1986), esp. VI (1), pp. 186–230; and the entry on L. Shih-Chen, the sixteenth-century pharmacist, in the *Dictionary of Scientific Biography* (New York, Charles Scribner's Sons, 1973), VII, pp. 396–9.

9 See, for example, B. Berlin, *Ethnobiological Classification: Principles of categorization of plants and animals in traditional societies* (Princeton, Princeton University Press, 1992).

10 For discussions of the metaphors of the tree and the bush, see S.J. Gould, *Eight Little Piggies: Reflections in natural history* (New York and London, Norton, 1993), pp. 113, 285–7 and 427–38.

11 For this theme in the Darwin succession, M. Kemp, 'Haeckel's hierarchies', in *Nature*, 395, 1 October 1998, p. 447.

12 B. Buchloh, 'The Museum Fictions of Marcel Broodthaers', in *Museums by Artists* (Toronto, Art Metropole, 1983), pp. 45–56, and D. Crimp, 'This is not a museum of art', in *Marcel Broodthaers*, exhibition catalogue (Minneapolis, Walker Art Center, 1989), pp. 71–81.

13 'Dix milles francs de récompense', 1974, published in English in the special Broodthaers issue of *October*, ed. Benjamin H. D. Buchloh, 42, Fall 1987, p. 46.

14 *Ibid.*

15 Foucault (1970), p. x, see note 4.

16 Marcel Broodthaers, 'Methode', in *Section des Figures (Der Adler vom Oligozän bis heute)*, vol. I (Düsseldorf, Städtische Kunsthalle, 1972), p. 13.

17 *Ibid.*

18 Ludo Bekkers, 'Gesprek met Marcel Broodthaers', in *Museumjournaal*, series 15 (2), April 1970, p. 66.

19 Michael Oppitz, 'Eagle/Pipe/Urinal', published in English in *October*, 42, Fall 1987, p. 154.

20 Marcel Broodthaers, 'Section des Figures', in *Section des Figures (Der Adler vom Oligozän bis heute)*, vol. II (Düsseldorf, Städtische Kunsthalle, 1972), p. 18.

21 Oppitz (1987), p. 155, see note 19.

22 *Ibid.*, p. 156.

23 Broodthaers (1972), p. 18, see note 20.

24 Susan Hiller, *At the Freud Museum*, Freud Museum, London, 1994.

25 Susan Hiller, 'Working through Objects', in *Thinking about Art: Conversations with Susan Hiller*, ed. Barbara Einzig (Manchester, Manchester University Press, 1996), p. 227.
26 Susan Hiller, *After the Freud Museum* (London, Book Works, 1995), n.p.
27 *Susan Hiller*, Tate Gallery, Liverpool, 20 January–17 March 1996.
28 Hiller (1996), p. 240, see note 25.
29 Hiller (1995), see note 26.
30 'It is not all really available for us any more', in Hiller (1996), p. 210, see note 25.
31 Hiller (1995), see note 26.
32 *Ibid.*
33 M. Righini Bonelli and T. Settle, *The Antique Instruments at the Museum of History of Science in Florence* (Florence, Arnaud, n.d.), pp. 16–17.
34 See *herman de vries: documents of a stream: the real works, 1970–92,* exhibition catalogue, ed. P. Nisbet (Edinburgh, Royal Botanic Garden, 1992).
35 *Ibid.*, introduction.
36 The British Museum installation 26 April–21 June 1997 was part of a series of artists' interventions in London Museums, shops etc., curated by Neil Cummings under the title *Collected.*
37 M. Fleming, *Atomism and Animism*, brochure to accompany exhibition at the Science Museum, London, 25 May–31 October 1999.
38 Both quotations *ibid.*
39 Cornelia Parker, *Cold Dark Matter: An exploded view*, exhibition catalogue, ed. A. Searle (London, Chisenhale Gallery, 1991).
40 M. Kemp, 'Mendeleev's Matrix', in *Nature*, 393, 11 June 1999, p. 527.

6

Creativity and the Making of Contemporary Art

Mike Page

> *… My brain's like*
> *the hive: constant little murmurs from its cells*
> *saying this is the way, this is the way to go.*
>
> JO SHAPCOTT, from 'The Mad Cow Talks Back'.[1]

This chapter looks at some ways in which mental processes are involved in the making and viewing of contemporary art, particularly in relation to an exhibition called *Thinking Aloud* curated by the British artist Richard Wentworth, which opened in 1998.[2] The exhibition included an eclectic array of objects. Among a range of contemporary artworks (none of them the artist's own) were interspersed items whose appearance in an art gallery would probably have surprised their creators. Befitting an exhibition whose original working title had been 'First Thoughts', the show contained many artefacts of the creative process, gleaned from the fields of engineering, architecture and design: a preliminary sketch of what became the Crystal Palace, hastily scribbled by Sir Joseph Paxton on a piece of blotting paper; Henry C. Beck's original sketch of the London Underground Railways Map, roughed out with an apparent insouciance, ignorant of the admiration which the design was later to enjoy; likewise a drawing of the Mini's subframe and suspension, by its designer Sir Alec Issigonis, and a 'butchered' vacuum cleaner, an early prototype of James Dyson's subsequently successful design. Wentworth also included copies of patent applications, such as that for the flip-top cigarette packet, necessarily more detailed, more formalised than the raw sketches described above, but loaded with the optimism implicit in a bid for ubiquity. Representing a still later stage in the evolution of an idea, with ubiquity now ensured by commercial success, were moulds for the industrial production of car tyres, bricks and condoms (the latter an ironic collision between the desire for repeated reproduction and, in some sense, its exact opposite).

More sinister inclusions were those resonant with associations of war: a pristine, white body bag, emphasising by its very flatness the grisly physicality of its anticipated contents; sticks of dynamite modelled in wood *(In Anticipation of an Explosion* by the artist Cornelia Parker); various objects in disruptive-pattern camouflage, either depicted in paint, or photographed, or, as in the case of a camouflaged skateboard, physically present in the gallery with a visual prominence entirely subversive of the camouflage's intended function; a number of preliminary sketches of Sir Edwin Lutyens's Whitehall Cenotaph; a plaster model of an explosion taken from a German toy soldier set, and a study in pencil of the real thing as witnessed by Keith Henderson in France, 1917; a wooden dummy rifle used as a training aid at Eton College in the early part of this century; and a doodle in red pencil on blotting paper, distinguished only by having been taken from the place occupied by Lloyd George at the Versailles Armistice meeting of 1918. Not unconnected was a collection of prosthetic limbs, their varying degrees of sophistication all too obviously failing to approach that of their natural counterparts. And alongside these were other abstractions of the natural world, notably maps, globes and surveys, some continuing the military theme, like maps for the D Day landings; others showing, for example, the sea areas of the shipping forecast, robbed of their somniferent charm, or the streets of London which comprise the 'Knowledge' of the London taxi driver.

Another 'thread' consisted of arrangements of materials and of objects, configurations shaped by unseen forces, such as those evidenced in patterns of soil erosion, photographed by Walker Evans in the Mississippi of the Depression era; or William Smith's nineteenth-century drawings of geological strata, laid down as sediment and later exposed by the grind of the elements; or the window display of a household supply store in Pittsburgh, also recorded by Walker Evans, showing a constellation of items as artfully organised as the exhibition itself, placed by an unseen hand subject to various unspoken constraints.

There were also a number of objects less clearly themed: a cardboard box full of remaindered toy dinosaurs; a bird's nest, with two of its sides perpendicular, evidence of the corner it once occupied; Simon Starling's low-grade, home-made replicas of a design classic, Charles Eames's 1955 DKS chair; and some other less deliberate 'failures', like an eighteenth-century mousetrap of questionable effectiveness; Keith Coventry's bronze rendition of kebab meat rotating on its spit; Bruce Nauman's screenprint depicting simply the words AH HA, the reflective symmetry emphasised by a light/dark colour scheme; and neuroscientist Tim Bliss's photograph of an activated synapse between two brain cells, reductively entitled *Unit of Thought*.

While, for descriptive convenience, the exhibited objects are listed here under thematic categories, in reality these threads were interwoven through the gallery space so as to foil just such classification. The propensity to classify was

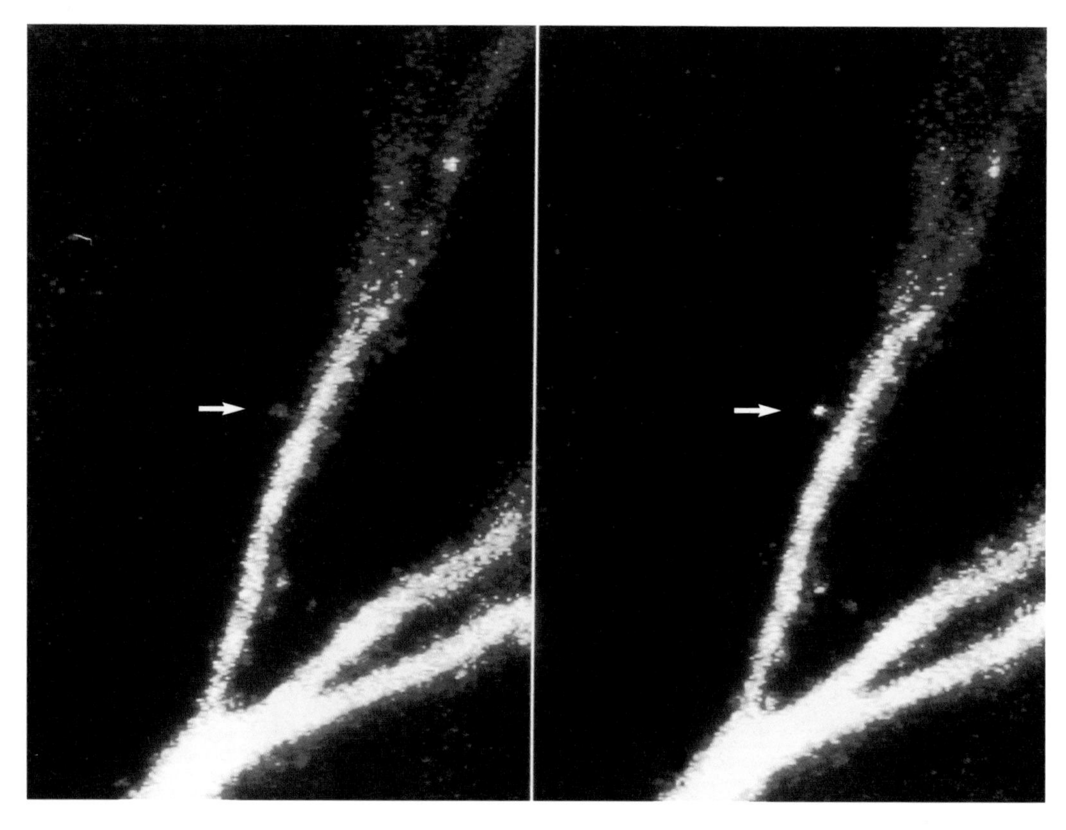

Unit of Thought. Two laser-scanned confocal microscopic images of a part of a cultured nerve cell from a mouse hippocampus. Courtesy of Drs Tim Bliss, Nigel Emptage and Alan Fine. Division of Neurophysiology, National Institute for Medical Research, London.

This pair of images shows a portion of the dendritic tree of a nerve cell in the hippocampus, a part of the brain involved in laying down autobiographical memories. The cell has been filled with a fluorescent dye which is sensitive to calcium (it becomes brighter with increased calcium concentration). The dendritic tree is studded with processes called dendritic spines (arrow, left), where axons from other nerve cells (not visible) make contacts, called synapses, with the imaged cell. There can be tens of thousands of such dendritic spines on a single nerve cell. When a nerve impulse arrives at a synapse, a chemical neurotransmitter is released from the axon terminal, and diffuses across the synaptic space to activate the dendritic spine. This produces a brief increase in calcium concentration which can be seen in the second image as a brightening of the spine (arrow, right). We are seeing a single synapse in action.

itself highlighted in the exhibition by a number of objects relating to, or resulting from, processes of classification: a poster depicting annotated images of a bewildering variety of breeds of cow; a set of toy farm animals appropriately labelled; reference objects for the units of a yard, a metre and a kilogram; the coded notebooks of Gaspard Marin (1883–1969), an obsessive taxonomist of everyday life; Simon Patterson's lithograph *The Great Bear*, a pastiche of Beck's Underground Map, with stations replaced by philosophers, artists, writers, and so on, connected by the coloured tracks of thematic relation, but defying neat compartmentalisation thanks to the map's familiar web of intersections; and, as a reminder of the lexical organisation of encyclopaedias and dictionaries, Joel Fisher's *Double Alphabet (1979–1981)*.

Reaction and recognition

The fact that we can make something meaningful of art, even of contemporary art with its calculated dislocations, is testament to the extraordinary flexibility of our mental processes. The principal of these is recognition. The recognition process is a complex and multi-levelled one, but let us first consider the primitive act of recognition as commonly understood. As we wander through the gallery, our gaze falls upon an object and, in a moment, it is recognised: 'Oh, it's a wooden rifle', 'It's an early Tube map', 'It's a kebab modelled in bronze'. Already, a considerable mental feat has been accomplished. Just imagine the number of possible objects that might have been recognised: a toy mouse; a silhouette of Africa; a hot dog modelled in pewter. The list of objects is effectively infinite, all recognisable with approximately equivalent ease within a couple of seconds. The magnitude of the achievement becomes even more evident when we consider that recognition proceeds smoothly, despite its operating on a stimulus that is, in one sense, bound to be novel. Just as Heraclitus asserted the impossibility of stepping into the same river twice, one can never perceive even something as paradigmatically familiar as the back of one's own hand in exactly the same way twice. A change in aspect, a subtle change in lighting, goose pimples, worsening myopia, retinal 'noise' – all will conspire to differentiate the two experiences. And to sharpen the comparison with Heraclitus' river, the brain that perceives the hand a second time will be slightly different from that which did so the first time. Thus every scene with which we are confronted is to some extent unfamiliar. As a consequence, the recognition of objects must proceed by generalising from similarities between the item to be recognised and those items familiar from our past experience. The similarity might be visual (e.g. of colour) or spatial (e.g. of shape) or it might be of smell, sound or feel – we might even choose deliberately to bias our notion of similarity towards one sensory dimension and away from others. Whatever our choice, our perceptions and cognitions are driven by similarity and thence on generalisation from our past experience.

Generalisation in recognition is so basic a property of our mental life that it is easy to forget the magnitude of the achievement it represents. One of the most salutary lessons in the recent history of cognitive psychology has been the difficulty of simulating on computers many of those perceptual and cognitive processes which we take for granted. In many ways the metaphor of the computer (as opposed to the tool itself) has proved a hindrance rather than a help in understanding the workings of our own brains. For example, the computer components biomorphically dubbed 'memories' (Random Access Memory, Read-Only Memory, etc.) have a radically different structure from the memories stored in our brains. Brain memories possess a property which cognitive psychologists call 'content addressability', which means that an environmental stimulus can access directly those memories stored in the brain which have related content. For example, if presented with a photograph of the Queen, you will achieve almost instant access to memories relating to her. This is not because the information in the photograph contains some index to the location of the memory in your brain as expressed in physical co-ordinates (e.g. 3 cm behind the eyes, 0.5 cm left of the midline) but rather because the information available in the picture relates to the pattern of connections leading from your eye to the higher areas of your brain dedicated to visual-pattern recognition and semantics. It is this relation that activates the appropriate areas of your brain so as to enable, say, rapid identification of a visual stimulus.

By contrast, if the same photograph were presented for 'recognition' to a conventional computer, no such direct access would be achieved. Neither the electronic address at which information is stored (e.g. bytes 14365 to 15437 in the third section of the second partition of hard drive C) nor the means by which that address is accessed, bear any meaningful relation to the *representational* content of that information. Thus the content of the stimulus (e.g. a digitised picture of the Queen) cannot help in the task of locating the relevant memory and the computer memory cannot, therefore, be described as content-addressable.

A content-addressable memory allows us to operate efficiently in the world and, of course, comes usefully into play when we look at art. Thanks to our flexible and fuzzy specification of 'content', we can, as noted earlier, focus on one dimension of a stimulus (such as shape) rather than another (smell) and generalise accordingly – recognising, for example, a bronze sculpture as a representation of a kebab. Of course, this process of recognition does not induce the belief that the sculpture actually is a kebab. Rather, it enables the identification of a particular dimension of the object, in this case its semantic referent. Moreover, recognition represents a gateway to a rich web of associations with the recognised object: when we last had a kebab, in what part of the world kebabs originated; what a kebab tastes and smells like.

So a content-addressable memory allows us to recall our knowledge of a recognised item; it allows us to recollect, literally to collect again, the information we possess regarding the object before us. Importantly for art, this information includes any emotional associations that we carry with us. Thus when we see the Eton rifle, we are reminded of the boys who later went 'over the top' shouldering the genuine article; when we see a map of the shipping forecast we are reminded, perhaps rather elliptically, of a warm bed; when we see a Tube map, we hear in our mind's ear the ominous intonation, 'Due to a signal failure at …', and dread the long trundle home. All these associations, these links, are

Keith Coventry, Kebab Sculpture no. 1, 1998, in the exhibition *Greeks* (1998). Bronze, 111 x 38 x 38 cm. Courtesy of the artist and Richard Salmon Gallery, London.

learned by experience. The associations are encoded in some of the million billion connections, known as synapses, that join together our 100 billion brain cells. It is because our knowledge is stored in the pattern of connections between cells that the study of such highly interconnected brain systems is called 'connectionism'.

Attractors and classification

Along with the ability to recognise and to generalise, comes the 'propensity to classify' alluded to earlier. Once we have seen a cow, and heard it labelled as such, our natural inclination is to attach that label to other animals that look similar. The label will build around it what is known in connectionist jargon as a 'basin of attraction'. The analogy is a gravitational one and refers to the behaviour of a ball placed at a peripheral position in a hemispherical basin. The ball will roll around, but will ultimately come to rest at the lowest point, the stable spot at which the forces acting on it, symmetrically upwards from the bowl and downwards from gravity, are in equilibrium. The connectionist theory maintains that familiar items are represented by similarly stable states of activation of the brain, that is, states towards which other less stable states are 'attracted'. If familiar objects, or the labels for them, are represented by these stable points, then related objects, unfamiliar but similar, will, if similar enough, be caught in the basin of attraction surrounding a given stable point and will be recognised and classified. In this way, the brain state elicited by a novel object that looks like a duck and quacks like a duck will probably get attracted towards the stable point representing the concept and the label 'duck'. Our brain will support many different stable states, each corresponding to some familiar concept or label. As we move around the world, our eye alights on a novel stimulus (perhaps the next exhibit in the show) and this establishes an unstable brain state. The metaphorical ball in the basin is rolling again, around a personalised landscape of attractor basins representing possible labels for this novel object. Providing the novel object is classifiable as an example of a previously learned category, the ball will find itself being attracted towards a local low point, like a hapless skier being drawn irresistibly towards the valley floor. As the stable point, the state of equilibrium, is approached, a sense of familiarity grows. As the process is completed, as the rolling ball comes to rest, recognition dawns and the symmetry of Bruce Nauman's AH HA moment is achieved.

Identifying themes

Another form of recognition will have become evident to anybody who witnessed Richard Wentworth's show, namely the recognition of thematic

Model of explosion, part of a toy soldier set, German 1930–9. Plaster composition. Courtesy of George Hardie. Photo: Mike Parsons.

structure. Objects in the exhibition were not placed thematically in the gallery space and a visitor to the gallery would encounter these items in an order not in itself conducive to the identification or recognition of themes. The effect is disquieting, particularly in the initial stages before reference points have been established. You wonder whether you will ever be able to deal with what seems at first like a stream of disconnected ideas. And yet, half way round the first gallery the 'parsing' of the objects becomes easier, less effortful, even a source of enjoyment and exhilaration. Experience with the items encountered earlier on establishes memories which can be 'addressed' by the content of later items, in the manner just discussed. There is no need for an exhaustive trawl through all the objects seen to date, looking for a match. Thus, when one comes across a sketch of the underside of a Mini, it addresses directly the memory of the scribbled Crystal Palace encountered five minutes previously. We have moved up a level: it is not simply that objects are being recognised for themselves; our minds almost compel an organisation into themes, each of which has been established spontaneously and without any explicit instruction, in the few minutes since we entered the exhibition.

Recognising rightness

And so to a third type of recognition – one that links with the 'creativity' of the title of this chapter – that is, the recognition of something's being 'just right'. From the perspective of cognitive psychology, this is perhaps the most mysterious of the processes discussed so far. From the point of view of the experiencer, however, it is the most familiar. In *The Story of Art* E.H. Gombrich gives the example of the non-expert arranging a bunch of flowers, placing a wayward bloom here, or a wilful twig there, before stepping back to survey the whole and to assess its 'just-rightness'. If it's just right, or right enough, then stop; otherwise back to the arranging.

But what does it mean to be 'just right'? It is clearly a subjective judgement – just right for you might not be the same as just right for me, though we can imagine circumstances in which some level of agreement would be likely. But knowing that the judgement is subjective doesn't help us to discover the underlying mental processes. The umbrella term under which such processes are often grouped is 'constraint satisfaction'. To give another example: suppose you are standing on a reasonably crowded Tube train (perhaps the same one as was delayed by a signal failure earlier). Where do you stand? Well, there are a number of constraints to be satisfied: near enough to the handrail to allow an effective grip in the event of heavy braking; as far as possible from that bloke with the guitar; not so close to the person standing next to you as to appear overly familiar, and so on. It may be that these constraints are explicit, as in the desire to avoid the imminent reprise of Nirvana's greatest hits, or implicit (i.e. not 'conscious'), as in the appreciation of fellow passengers' need for personal space. But if all goes well the position you finally adopt will be a satisfactory resolution of all these constraints, consistent with the more fixed physical constraints of the carriage. Like the ball in the bowl you will come to rest at the point at which the forces or constraints acting on you are in equilibrium.

The reason that understanding the precise mechanisms of constraint satisfaction is difficult is partly because the variety of constraints that can operate (physical, social, cultural etc.), combined with the number of possible situations in which they might do so, makes for an unappealingly large number of options for study. Scientists favour a narrower focus. Nonetheless, some connectionist models have been shown to demonstrate constraint satisfaction, at least in general terms. Put simply, constraints are represented 'neurally', that is, a brain cell or group of cells is simulated so that the group is assumed to be activated in order to signal the presence of a particular constraint. The neural representations of constraints that are mutually consistent are connected by positive, mutually supporting connections; those of opposing constraints are connected by negative or competitive connections that discourage their simultaneous activation. By allowing the system to evolve dynamically the pattern of activation across all the groups of cells, a stable point can be reached at which the maximum number of constraints is maximally satisfied. Similarly, as we move our blooms around in the vase, we do so in an attempt to raise the 'harmony' of the whole with reference to the network of perhaps implicit constraints represented in our synaptic store.

Many items in *Thinking Aloud* bore witness to the unseen guidance of implicit and explicit constraint. As noted earlier, the way in which objects were arranged in the window of an American household store said much about the constraints operating on the unseen shopkeeper who had placed them there. The eerie patterns of soil erosion recorded by the same photographer spoke eloquently of the forces to which the land had been subjected. The perpendicular bird's nest tells us much about the physical constraints that shaped its construction. Richard Wentworth has often spoken about the ways in which objects around us come to be arranged, from a glass of water placed on a table to the motley arrangement of furniture outside his local junk shop. His own work reflects this fascination and he has described his experience of being under implicit constraints as he arranged this exhibition in each of the three gallery spaces in which it appeared. Had he been asked to write a guidebook giving explicit instruction on how to go about arranging his collection in a given space, one guesses he would have found it difficult, if not impossible, to put his intuitions into words. Nevertheless one may assume that for him the arrangement ended up, like the flowers in the vase, somehow 'just right'. As Wentworth demonstrates, our assessment of quality need not be explicit. Implicit knowledge can play a very significant role in our psychological life. We should not, however, confuse this implicitness with innateness. Beyond some

Walker Evans, *Household Supply Store, Bethlehem PA, November 1935*. Photograph. Courtesy of the Library of Congress, Washington DC.

possibly innate preferences, such as a liking for symmetry or for certain colour combinations, it is probable that the bulk of what informs our sense of rightness is acquired via extended interaction with our environment and all the cultural and social values it embodies.

Know-how and know-whether in creativity

This ability to assess the 'rightness' of something without necessarily being able either to specify what makes it right or, indeed, to generate something else right, highlights the distinction between two distinct types of knowledge: *knowledge-how* (more colloquially 'know-how') and what the physicist Mike Greenhough has termed *knowledge-whether*.[3] Knowledge-whether is the sort of knowledge that one exhibits when one expresses a judgement about something. Importantly, it does not presume knowledge-how: you might not be able to hold a note but you might still be able to tell a well sung *Lied* from one badly sung; or you might have the proverbial two left feet, yet be able to appreciate the finer points of the free-kick curled around the wall; or be unable to design a good, clear map of the London Underground, yet be able to spot one when you see it. The logical relationship between the two knowledge types is not symmetric; knowledge-how is likely to imply knowledge-whether. It would be extraordinary, for example, to find someone who had enough ability to win at Wimbledon but who had insufficient judgement to tell a fine tennis player from a mediocre one.

This distinction between know-how and know-whether introduces the possibility of two general creative strategies. To illustrate this, let us suppose you were asked to produce an attractive layout for a garden. One strategy might be to read books and take courses until you had acquired sufficient know-how to permit a direct and explicit statement of a suitable design. Another strategy might be to rely on previously acquired good taste in such matters, that is, your instinct as to whether a garden looks nice or not, and simply to arrange and rearrange horticultural features until you arrive at a configuration that looks just right. You might even get hold of a computer program which can generate a whole host of possible designs, all fitting some fairly general external constraints of, for example, size and cost. Your only remaining job would be to pick, from the multitude of plans presented to you, the one that pleases you best. Using this method, which involves no explicit knowledge at all, a perfectly satisfactory design might be realised. Indeed, given enough time, it will be realised. To be sure, the process of actually designing the garden might be slower with the know-whether strategy than with the know-how strategy. One could, however, envisage a computer program which would allow you to preserve the best features of a less-than-ideal design – those which you had considered to be 'the right sort of thing', limiting its subsequent suggestions to the generation of variations on the specified theme. Guided by such a program, it is conceivable that a creative strategy based only on the repeated expression of judgement might be preferable to, or at least more achievable than, a strategy requiring direct and explicit statement of a solution.

Generate-and-test in creativity ...

So to what extent does creativity in the arts, sciences, engineering and design rely variously on processes of know-how and know-whether? It is likely that most instances of creativity will involve elements of both. The doodles, maquettes, studies, sketches and prototypes that populated *Thinking Aloud* gave ample evidence of the trial and error nature of the creative process. They bear witness more generally to a strategy of *iterated* (i.e. cyclically repeated) *generate-and-test* (a form of words borrowed from the philosopher Daniel Dennett), a strategy general enough to encompass elements of both explicit statement and implicit preference. Within the framework of generate-and-test, we might ask several questions of the arts, the sciences and the crafts (including engineering and design). Who does the generating? And what characterises the generative process? Is it largely based on explicit knowledge or does it contain a significant element of random generation? Focusing here on the visual arts, the answers will surely depend on the artist and the artefact in question. A painter of the early Middle Ages, commissioned to paint an altarpiece dedicated to the Virgin Mary, would come to the task armed with established rules for the painting and positioning of the relevant cast of characters, yet would probably still have generated a number of preliminary studies, to try out particular arrangements before settling on one that seemed 'right'. In other less rigidly prescribed cases, the subject of a painting might 'emerge' from a process in which deliberate or conscious direction has been intentionally minimalised. Take, for example, Gombrich's paraphrase of Klee's description of his artistic method:

> Klee tells us how he began by relating lines, shades and colours to each other, adding a stress here, removing a weight there, to achieve the feeling of balance or 'rightness' after which every artist strives. He described how the forms emerging under his hands gradually suggested some real or fantastic subject to his imagination and how he followed these hints when he felt he would help and not hinder his harmonies by completing the image he had 'found'. It was his conviction that this way of creating images was more 'true to nature' than any slavish copy could ever be ... it is the same mysterious power that formed the weird shapes of the prehistoric animals, and the fantastic fairyland of the deep sea fauna, which is still active in the artist's mind and makes his creatures grow. This method again recalls our doodles, when we let ourselves be surprised by the outcome of our idle pen games, except that to the artist this can become a serious matter.'[4]

Clearly Klee recognised the 'mysterious power' of largely undirected generate-and-test (though he would doubtless have called it something different). Moreover, he saw in it a reflection of the creative work of nature, an observation to which we shall return shortly.

And what is the nature of the test process in the visual arts? Again, a general answer is not particularly helpful. There are artists whose motivation, as far as we can tell, is not to please others, but rather to satisfy themselves. Paul Cézanne, for instance, strove to redefine the painting of form and of light, but was apparently uninterested in the opinion of either the critical establishment of the time or of a wider public. We might hazard that the tests inherent in his productive cycle were based on a personal assessment of whether the artefact before him solved the problems with which he was troubled. Other artists, in particular those associated with wider 'movements', might have been more disposed to value the judgement of their peers, those that shared their credo. But unlike scientists, artists are at liberty to choose their own criteria to decide what constitutes sufficient success for a work to be considered complete.

... And creation

Finally, we come to creation – or Creation? Richard Wentworth's exhibition was the product of a creative act. It was also full of the artefacts of creativity. It also made many allusions to Creation itself (though perhaps not as traditionally conceived). If you ever doubt that a process of blind know-whether is capable, given enough time, of producing the most beautifully designed artefacts, then simply look in the mirror. The design of the human body, and that of most other living things, is the result of a process of evolution in which effectively random generation has been allied with the pure 'opinion' of an environmental test. The generative step, the genetic mutation or 'mis-copy', has no direction. It is not teleological, that is, it is not aligned towards the achievement of some ultimate goal. And the testing stage is embodied only in the effect of an unsentimental environment on the differential ability of a 'prototype' to reproduce. To paraphrase the art cliché: the environment knows nothing about designing creatures, but it knows what it likes. As is characteristic of such processes, the design process is a slow one. The process takes advantage of cumulative incremental change, one generation being made up of only slight variations on the previous one, rather than being a radical, and therefore probably unviable, redesign of it. Like the superior garden-design program, it thus locates more viable solutions more quickly than it might otherwise have done. More quickly, maybe, but still mind-numbingly slowly. But that's the beauty of non-teleological processes: when you've got nowhere in particular to go, it doesn't really matter how long you take to get there.

The reading of a work of art is a personal matter. One might view Richard Wentworth's exhibition as an essay on Evolution. After all, what more salutary reminder could one have of the vicissitudes of natural selection than a box of remaindered dinosaurs? And who could resist seeing in the model of an exploding shell an image of the explosion that may have hastened their demise? Explosions release energy and energy can rearrange things. Explosions can rearrange physical landscapes, gouging craters into battlefields and planetary surfaces alike; they can rearrange political landscapes, as Lloyd George could attest as he doodled in Versailles; and they can speed the rearrangement of evolutionary landscapes, leaving the once-dominant dinosaurs high and dry.

And what better illustration could one have of the powers of evolutionary 'judgement' as against those of the human intellect, than by comparing our own elegantly functioning limbs with their clumsy, reverse-engineered prosthetic counterparts, or by comparing the crude regularity of disruptive-pattern camouflage with the finely wrought delicacy of its natural inspiration. And when we admire William Smith's pre-Darwinian sketches of geological strata, our post-Darwinian minds are turned to the crucial evidential power of their fossilised contents. Like *Thinking Aloud*, the strata contain a selective record of early prototypes, all successful for a while (i.e. long enough for them to have been fossilised), some representing links in a chain ending in extinction and others representing direct ancestors of earth's current inhabitants. Wentworth has often referred to his show as anti-heroic, a celebration of the mundane, the good-enough. In the grind of natural selection, it is the good-enough that survive, for the concept of the best is meaningless. There can be no such thing as a perfectly adapted creature in an environment that is always in a state of flux. Indeed, the concept of the 'perfect' is an anathema to natural selection. It is imperfection that drives evolution – without variation due to genetic mutation or miscopying, there is no heritable differentiation on which the environment can work. Like Simon Starling's Eames chairs, imperfect replication should be celebrated. Without it, we would not exist.

The evolution of creatures and the evolution of ideas share some common characteristics, particularly inasmuch as they are both examples of generate-and-test. The similarities have been noted many times before but we should be wary of overloading the analogy. The educational psychologist David Perkins has detailed more formal distinctions to be drawn between biological evolution and human invention.[5] Nature can only 'reason' in the form of physically embodied creatures while inventors can imagine things that might never exist in the domain of their intended solution. Similarly, in natural evolution nothing can be learned from a design which fails, while in invention failures can be as eloquent as successes. Natural evolution is effectively guaranteed to be slow and gradual whereas the inventive process can involve leaps into previously uncharted territory (perhaps, as Perkins notes, like Alexander Fleming, propelled in an entirely new direction by the witnessing of a chance encounter of mould and bacterial culture). Artists and other inventors can explicitly foster a creative frame of mind, deliberately managing the degree to which they continue to work on existing ideas instead of throwing themselves open to a new line of thinking, whereas no such meta-theoretical management is available in the evolution of biological kinds. And, unlike biological evolution, invention can take advantage of meta-ideas, abstracted from a variety of specific applications: Perkins gives the example of the meta-idea 'put something with its inverse' to explain such artefacts as the claw-hammer, which can insert and remove nails, or the eraser-tipped pencil, which can draw or obliterate.

Thinking aloud

Finally, it is worth raising the question why the intermediate products of the creative cycle so often take on a separate existence as artefacts (sometimes available for our later perusal), from the engineer's back-of-an-envelope to the sculptor's maquette. The answer is partly explained by the limitations of the human brain. So far as we know, we are unique among creatures in the degree to which we are able to entertain what psychologists call 'counterfactuals'. In other words, we can imagine things to be different from the way in which they are. We can imagine what it would be like if our bathroom were painted yellow, or if the shops were open 24 hours a day; we can plan our future life based on the expectation of a growing family; or we can envisage a car that flies. This fluency with counterfactuals can sometimes be as much of a curse as a blessing, for it can be a source of disappointment, guilt, or envy. It can also lie at the root of damaging mental disorders. Indeed it is likely that many of the most prevalent mental diseases, like depression, anxiety or post-traumatic stress disorder, have as an important component the sufferer's ability to posit how things might be or might have been. Nevertheless, it is clear that without our talent for counterfactuals, our levels of creativity would be drastically reduced, perhaps to nothing.

Yet it is the degree to which our conceptual imagination runs ahead of our ability to visualise that must partly explain the abundance of creative artefacts, such as those which have found their way into *Thinking Aloud*. For on occasion the constraints placed on us by our limited memory capacity, our limited capacity to entertain several thoughts at once, force us not just to think aloud but also to think in objects. In a memorable phrase, Wentworth referred to Dyson's mangled vacuum cleaner as being not just for cleaning the floor but also for cleaning the mind. There comes a point where the best way of determining what would happen if I connected this thing to that, or did that bit in blue, is actually to do it. Maybe it is not necessary to do it at full scale or in fine detail, but just to externalise enough of the idea to permit the exercise of tasteful assessment and to free up some valuable brain resources for other things.

Not only can we think aloud and think *in* objects, but to a large extent we can think *with* objects. As Daniel Dennett has persuasively argued, our creative powers have been immeasurably boosted by the panoply of what he calls mind-tools that we have gathered around us. The alphabet and other scripts, musical and mathematical notations, pliers, the scientific method, shopping lists, word processing programs and other computer software, fax machines, the calendar, visual art, tape recorders, and so on, and so on: all are mind-tools. It is this accretion of mind-tools, rather than a fundamental change in our biological make-up, that makes us more creative than our Stone Age ancestors. We have learned to build our intelligence into the world around us and this intelligence

Cardboard box of remaindered toy dinosaurs, 1998. Photo: Mike Parsons.

is, unlike that embodied in our synapses, cumulative across generations. We are in a virtuous circle in which more sophisticated artefacts of intelligence 'breed' yet more of the same. By externalising our knowledge we have freed our minds for further inventive pursuits. It is these inventive pursuits that were both exemplified and recorded by Richard Wentworth's exhibition.

Notes and references

All items illustrated in this chapter are from *Thinking Aloud* (1998–9), a national touring exhibition, curated by Richard Wentworth.

1 Jo Shapcott, from 'The Mad Cow Talks Back', in *Phrase Book* (Oxford, Oxford University Press, 1992).

2 *Thinking Aloud* (1998–9), at Kettle's Yard, Cambridge, the Cornerhouse, Manchester, and Camden Arts Centre, London, was organised by the Hayward Gallery, London, for the Arts Council of England.

3 Mike Greenhough, 'The Muse and the Machine', in *Mathematics and Music*, ed. J. Fauvel, R. Flood and R. Wilson (Oxford, Oxford University Press, in press).

4 E.H. Gombrich, *The Story of Art*, 16th edn (London, Phaidon Press, 1995), p. 578.

5 David N. Perkins, 'Creativity: Beyond the Darwinian Paradigm', in *Dimensions of Creativity*, ed. Margaret A. Boden (Cambridge, MA, MIT Press, 1994).

7 Uncertain Entanglements

Richard Bright

I like the face of this
theoretical physicist
which appeared
an abstraction
from an unknown and undefinable
totality
and has vanished
leaving us a theory –
a theatre –
in which we sense the whole

JEREMY HOOKER, from 'Workpoints'.[1]

Simple or complex? – art and physics

The eminent physicist Richard Feynman once remarked that there is an inherent problem with the apparently simple statement 'magnetism is like a spring' in that it requires much background knowledge in order to understand fully the relationship between the one and the other. One must first need to know what a spring is. One must know what tension means and understand terms like stretching and compression, forces, stored energy and so on. One would also have to know what a magnet is and to understand magnetic fields and dipoles, electromagnetism, moving charged particles and electrons.

The same could be said about a hypothetical consideration of the relationship between art and physics. A comprehensive understanding of each is essential before one begins. Superficially, one could claim that both art and physics have some similar areas of interest. Both explore the physical nature of materials but their reasons for doing so are different. Physics analyses; art makes, or manipulates in unusual ways. In respect of contemporary physics, however – quantum theory, relativity and cosmology – there are potentially close connections because both are concerned with questions about the ultimate nature of reality. Both are concerned with how we see ourselves in relation to nature, whether as objective observers or as subjective participants. Indeed, in modern times, the distinctions between notions of 'subject' and 'object' have become blurred in each.

Consider the following statement, also by Richard Feynman. Apparently reductionist at first sight, hidden within it is the whole spectrum of knowledge about nature, from the simple to the very complex:

> If, in some cataclysm, all of scientific knowledge were to be destroyed, and only one sentence passes on to the next generation of creatures, what statement would contain the most information in the fewest words? I believe it is the atomic hypothesis – that all things are made of atoms – little particles that move around in perpetual motion, attracting each other when they are a little distance apart, but repelling upon being squeezed into one another ... and that everything that animals do, atoms do. There is nothing that living things do that cannot be understood from the point of view that they are made of atoms, acting according to the laws of physics.[2]

Is the world so simple? Or is it, in fact, extremely complicated? According to the cosmologist John D. Barrow, asking this question of a wide range of people would very likely result in a wide range of answers, with the bias veering towards the complicated. A particle physicist or a cosmologist would probably describe the world as wonderfully simple, 'elegant' and 'symmetrical', if only you looked at it in the right way.[3] According to these physicists, the whole of the universe conforms to a few laws of nature which govern the workings of just four natural forces – electromagnetism, the strong (nuclear) force, the weak (radioactive) force, and gravity – and that maybe, if you came back in a few years' time, there would be only one force, the Grand Unified Theory which you could probably fit onto the front of a T-shirt.[4]

However, a biologist, psychologist, economist or artist would almost certainly say that the world is a difficult mess of complex interactions. Besides citing examples from their own disciplines – the intricate structure and behaviour of cells, the unpredictability of human emotions, the tangle of interactions in money markets, fleeting effects of light, colour and shade in a landscape – they could also draw on many examples from daily life, from the changeability of the weather and the frustration of unexpected traffic jams, to natural evolution and the complexity of all living things, to say nothing of the myriad psychological nuances and problems which provide the richness of human life. So who is right? Why is there such a dichotomy between these two points of view? For John Barrow, the difference in attitudes 'betrays a difference in the nature of their subjects which results in a different perspective on the world'.[5] It is a difference which requires an understanding of the distinction between the laws of nature and their outcomes. In creative processes this involves symmetry-breaking.

Symmetry occurs if the same theme is repeated two or more times – in the patterns in snowflakes, in repeated musical rhythms, in wallpaper designs, in mirrors or the left/right-hand sides of human and animal bodies. If an object has symmetry, by its very nature it contains less information and only the basic theme and the way it is repeated is necessary to describe it. The physicists' laws of nature are fundamentally symmetrical – they remain the same at all places and at all times. However, the 'outcomes' of the laws of nature are asymmetrical. Very near the beginning of the universe, some 15 billion years ago at the 'singularity' of the Big Bang, which physicists refer to as 'zero time', all the forces of nature had equal strength and the masses of the elementary particles were identical, existing in pairs of matter and anti-matter particles. At approximately one-hundredth of a second after zero time, slightly more particles were produced than anti-particles, due to a tiny asymmetry in the way these particles interacted. By three minutes and 46 seconds after zero time, the process of energy conversion was complete to lay the foundation for the eventual formation of stars and galaxies. Our Sun was formed much later, about 5 billion years ago, from the debris of stellar explosions. Over the evolution of time, simple molecular forms burgeoned into complex systems of life on this planet.

Artists are intuitively drawn to complex systems but only a few have engaged with the scientific explanations and experiments which seek to understand the nature of formations. One such is the artist Susan Derges. In 1993 she produced a series of works entitled *Hermetica* devoted to the idea of making sound visible. These were not meant to present literal illustrations of scientific phenomena but were metaphorical hypotheses about what the invisible world might contain.[6]

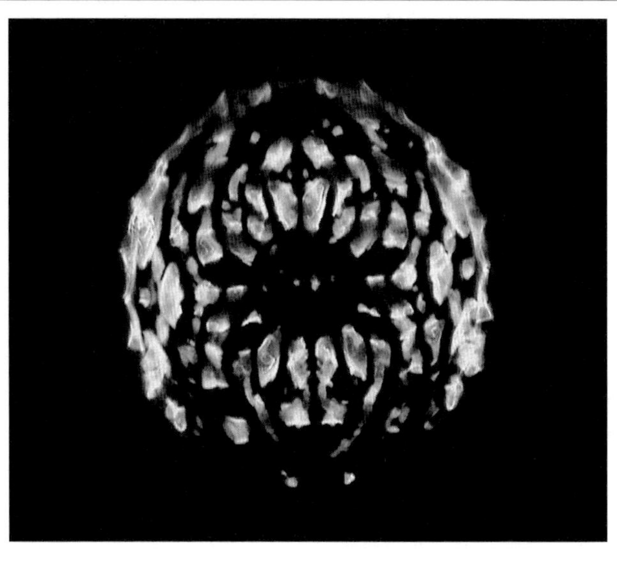

Susan Derges, video still from *Hermetica*, 1993. © Susan Derges. Courtesy of the artist and Michael Hue-Williams Fine Art Limited, London.

Hermetica was created by resting a very small drop of mercury in the base of a loudspeaker which emitted a sound of increasing frequency. The patterns formed by the mercury in response to the changing sound were recorded by a video camera. As the frequency became higher, the patterns became more complex, forming Chladni vibration patterns. As long ago as 1787, scientist and amateur musician Chladni provided in his book *Discovery of the Truth of Pitch* the first general treatise on the science of acoustics. In his experiments he covered metal plates with thin layers of sand, which were then set into vibration by drawing a violin bow across various points on the plates' edges. Through analysing the resulting sand patterns, Chladni classified the range of harmonics according to the geometric shapes produced by the vibrating sand. What is arresting about the images produced in Derges's *Hermetica*, because it is work to be viewed on video in real time, not as a series of static images, is the interaction between the simple and the complex, or, to use the correct scientific terms, between *order* and *chaos*. Certain frequencies produce symmetrical patterns – complex polygonal shapes – while others produce chaotic patterns. The most interesting pattern, both visually and conceptually, occurs at the point where the symmetrical meets the chaotic. Scientists refer to this point as *dynamic equilibrium* or the *edge of chaos*, a critical event where physical processes display the greatest complexity and are extremely sensitive to small changes. When viewing *Hermetica*, one feels that time has slowed down and that the evolving forms are delicately poised between order and chaos, exhibiting their potential for both. The simple and the complex seem to exist at the same time.

In another group of works, entitled *The Observer and the Observed* (1991), Derges broke up a jet of water, using a strobe light, to produce a standing wave form which, in photographs, appears as a series of droplets suspended in space in front of a seemingly unfocused image of her own face (see cover image). What is remarkable is that in each of the droplets the artist's face appears in sharp focus when the viewer is close to the artwork. If the viewer stands back, however, the face in the background suddenly comes into focus, while the droplet images become blurred. The face no longer behaves as a passive backdrop but shows the artist as a participator who, like the spectator, witnesses the event taking place in front of her eyes.

A characteristic of Derges's work is the setting up of specific situations in order to let the artwork 'make itself'. She sends sound vibrations through photographic paper and records the patterns produced, she uses strobe lights to capture arrested images and she makes 'cameraless' works by shining light at different exposures through beehives or rivers on to photosensitive paper, so that the ceaseless natural action occurring within each environment records itself in the very process of change. In this way she tries to minimise her personal manipulation of the outcome. The observer, whether artist or viewer, actually becomes a part of what is being observed and is not detached from the action. In this she draws, both intellectually and playfully, on an important theoretical conundrum in contemporary physics.

From the window to the observer: Newton to Einstein

'Observer participancy gives rise to information, and information gives rise to physics.'[7]

The mechanistic world view underlying classical Newtonian physics maintains that the laws of nature are absolute, are fundamentally deterministic and are universal. All the laws of classical mechanics are understood to be true in relation to what are known as *inertial frames of reference*. These may be visualised as a three-dimensional grid which permits the spatial fixation of any event, with a 'clock' recording the time of its occurrence. Within inertial frames of reference, time, space and motion are absolute. All the information which can be recorded about the motion of bodies on earth and in the universe is encapsulated in Newton's Three Laws of Motion.[8] If these are applied together with Newton's Law of Gravitation, a vast range of phenomena in the recognisable macro-world can be explained with great accuracy, from the paths of falling bodies on earth to the motion of the planets and including the trajectory of the *Voyager* space craft hurtling past Saturn. These all depend solely on Newtonian Laws.

Newtonian physics takes a particular perspective, almost in the literal sense of the word. It is posited on the belief that you can place the whole of nature in front of you and, looking at it through the 'window of objectivity', expose it to your intellectual gaze. This approach has a direct connection with the development of perspective in art during the Renaissance. Perspective in art was calculated geometrically to approximate the eye's projection from a particular stationary spatial and temporal point. It still predominates in representational art in the West and we accept it as realistic almost without thinking. Indeed, centralised perspective is more than an optical imposition on the world. It turns the viewer into pre-eminent subject and gives the illusion that his viewpoint is fixed and exact, his observation of nature, whether as artist or scientist, authoritative and 'true'. And although the image of the world is deliberately constructed to appear to be seen from the viewpoint of one particular (though invisible) observer, it appears to exist like that independently.[9]

Similarly, in the Newtonian paradigm, physical reality obeys its own laws, independent of any human observer, indeed, independent of whether there *are* observers at all. Things move and events happen as part of a chain of cause and effect. If something is acted on by a force, or communicated by a signal, it responds accordingly.

But at the turn of the twentieth century the Newtonian paradigm started to collapse. Scientists who, only years before, had declaimed like Lord Kelvin that 'all the interesting work has been done', were now faced with empirical data from experiments for which there were no adequate Newtonian theoretical explanations.[10] As a result of attempting to understand these discrepancies, the first quarter of the twentieth century saw momentous changes in the understanding of physics.

In ten remarkable years, Einstein transformed our understanding of space and time, of mass and energy, and of acceleration and gravity. In his Special Theory of Relativity, space and time are no longer considered absolute, but depend on the speed of observers moving relative to one another in different inertial frames of reference.

In Newtonian mechanics space is regarded as independent of time. In the Special Theory of Relativity, however, space and time become an inseparable part of *space-time*, a four-dimensional continuum in which *events* occur. Furthermore, both space and time are dependent on the observer's choice in the inertial frame of reference – moving clocks run slow, moving rods get shorter and the mass of moving objects increases – which gets more pronounced the more the speed approaches the speed of light. The famous equation $E=mc^2$ is a consequence of the Special Theory, which relates the interchangeability of energy and mass. For example, an object (such as an electron) moving at 99 per cent of the speed of light has seven times as much mass as when it is at rest. Only the speed of light remains a constant throughout the universe, irrespective of movements in space and time.[11]

The discovery of non-Euclidean geometry in the late nineteenth century, known as Riemann geometry after the German mathematician Bernhard Riemann, had a profound effect on the thinking of Einstein.[12] Euclidean geometry, sometimes called 'flat-space' geometry, is what we still learn at school with its straight lines, circles, and triangles in which the three angles always add up to 180 degrees and in which parallel lines never meet. By contrast, Riemann geometry describes 'curved' space, obtained by drawing a triangle on the surface of a sphere, with the angles adding up to more than 180 degrees.[13]

For Einstein this geometry formed the basis for a new theory of gravity, the General Theory of Relativity, which, following on from the features described by the Special Theory of Relativity, involves the relationship between mass-energy and the structure of space-time. In this theory, mass-energy produces 'curvature' in space-time. For example, in the vicinity of massive bodies like the Sun, space-time is curved, with the predictable effect that light from distant stars will be deflected as they pass through the curvature.[14] The evidence of light deflected from distant stars, seen during the solar eclipse of 1919, bore out the accuracy of Einstein's calculations and won him international acclaim.

In art, the essence of non-Euclidean geometry has been captured pictorially in the strange and confounding two-dimensional constructions created by Maurits Escher (1878–1972) who studied mathematics and physics and consciously addressed the problem of visualising the new theories. Escher's

M.C. Escher, *Relativity*, 1953. Lithograph. © 1999 Cordon Art B.V. - Baarn - Holland.

work encompasses two broad mathematical areas: the geometry of space, including perspective, and what we may call the 'logic' of space. Escher understood that the geometry of space actually determines its logic, and conversely, the logic of space determines its geometry. The logic of space involves those spatial relations between physical objects necessary for a plausible visual understanding of them – for example, no two objects can occupy the same space simultaneously. When the normal rules of perspective are violated the result is a visual paradox, more commonly described as an optical illusion.

In Escher's lithograph *Relativity*, three forces of gravity operate perpendicular to each other. It is impossible for the inhabitants of the different worlds represented in this picture to walk or stand on the same floor, because they have different conceptions of what is horizontal and what is vertical. Yet they share the same staircase. On the top staircase in the picture, two people are moving side by side in the same direction, yet one of them is going downstairs while the other is going up. Contact between them is out of the question because they live in totally different frames of reference and can have no knowledge of each other's existence.

M.C. Escher, *Circle Limit III*, 1959. Woodcut printed from five blocks.

Among the most important of Escher's works from a mathematical point of view are those dealing with the nature of space itself. The woodcut *Circle Limit III* represents hyperbolic space. This is one of two kinds of non-Euclidean space. To get a sense of what this space is like, imagine that you are actually in the picture itself. As you walk from the centre of the picture towards its edge you will shrink, just as the fish in the picture do, so that to reach the edge you have to walk a distance which would appear infinite. Indeed, if you were inside this hyperbolic space, it would not be immediately obvious that there was anything unusual about it. However, if you were a careful observer, you might begin to notice some odd things – all similar triangles are the same size; it would be impossible to draw a straight-sided figure with four right angles; this space doesn't have any squares or rectangles.

Escher's witty optical illusions demonstrate his intuitive grasp of spatial anomalies and have long been regarded with respect by scientists who, while they operate mainly through theoretical speculation and mathematics, still need to be able to visualise phenomena. The image in *Circle Limit III* has been used by cosmologists as a visual explanation of what theoretically happens at the boundary of the universe. In fact, Escher's images have been used as visual metaphors for a number of scientific theories, from perceptual psychology to quantum mechanics.

Quantum events

'I think I can safely say that nobody understands quantum mechanics ... Do not keep saying to yourself, if you can possibly avoid it, "But how can it be like that?" because you will go down the drain into a blind alley from which nobody has yet escaped. Nobody knows how it can be like that.'[15]

Elementary quantum mechanics was developed between 1900 and 1930 through the work of six men – Albert Einstein, Niels Bohr, Erwin Schrödinger, Werner Heisenberg, Paul Dirac and Max Planck – and it provided a startling new way of looking at the world. The central assumption underlying quantum theory, which comes into operation at subatomic levels, is that the objects and events are unpredictable and do not operate as they do in the macro-world. In particular, the notion of a well-defined trajectory has to be given up, and instead we can only contemplate probabilities. For example, Newtonian classical mechanics says that a particle will follow a single trajectory from A to B. In quantum mechanics, the natural interpretation of the mathematics is that the 'quantum entity' follows *every* possible path from A to B, but the different quantum paths interact in such a way as to cancel each other out everywhere except near the classical trajectory.

Ordinarily, we regard separate objects as independent of one another; indeed, arguably that is what the word 'separate' means. They exist and behave on their own terms and anything linking them together has to be connected to some tangible physical mechanism or force. However, in the quantum world, objects cease to be independent things, and can only be described in *relation* to one another. As Lee Smolin writes in *The Life of the Cosmos*: 'If the properties of any one particle are determined by its interaction with all the others, while that particle itself participates equally in its determination with all those others, then the laws of physics are a kind of system in which the influence of any one particle on the others feeds back to effect its own properties.'[16]

A key element of this is what is technically known as *entanglement*. One effect of entanglement is that if a measurement is made on one particular particle in any correlated pair, it is 'as if' the measurement has an immediate effect on the other particle, even if it is at the other end of the universe.

This is related to the strange phenomenon concerning notions about the state of a system before any observations or measurements are made. In the standard interpretation of quantum theory – commonly known as the 'Copenhagen Interpretation' – the state of a system does not describe 'the way things are' (as would be the case in classical physics); instead it describes only the *potentiality* for various results to be obtained *if* a measurement is made.

The *actual* occurs when a quantum system is observed or measured. Thus all we can know about 'reality' is what we measure with our instruments: the theory itself predicts only the probability that a particular experiment will come up with a particular result. As the physicist and information theorist Roy Frieden concisely states: 'Through the very act of observing, we actually define the physics of the thing measured.'[17]

This is a mind-defying notion worthy of epistemological contemplation and it is hard to recognise that it can still be suitably subjected to the rigours of scientific methodology and calculation. In 1935, the physicist Erwin Schrödinger proposed an amusing illustration of the ambivalence inherent in observing particles on a quantum mechanical level:

> Suppose we put a cat in a box with a radioactive atom, a Geiger counter, a hammer, and a poison bottle; further suppose that the atom in the cage has a half-life of one hour, a fifty-fifty chance of decaying within the hour. If the atom decays, the Geiger counter will tick; the triggering of the counter will activate the hammer, which will break the poison bottle, which will kill the cat. If the atom doesn't decay, none of the above things happen, and the cat will be alive. Now the question, What is the state of the cat after the hour?[18]

The most logical solution would be to wait an hour, open the box and see if the cat is still alive. However, if you consider the problem from a quantum perspective, then before the box is opened, the cat is in a state which can be thought of as both alive *and* dead (Schrödinger described the cat as being 'smeared out in equal parts' between the living and the dead). The act of opening the box will determine the one or the other. Just like the cat poised forever between life and death, quantum entities exist in a range of multiple possibilities, all equally possible and some maybe even incompatible. However, according to Chris Isham, Professor of Theoretical Physics at Imperial College, London, the situation is even 'worse' than this, since Schrödinger's statement of the situation is not quite correct. Specifically: the cat and the apparatus are not separate entities – they are entangled – and hence the cat alone is not actually in *any* state at all; it is neither alive nor dead!

Another central preoccupation of quantum physics concerns how the basic elements of matter should be regarded. Depending on the experiment, a quantum entity, for example light or electrons, may exhibit either a particle-like aspect or a wave-like aspect. A particle is usually regarded as having a precise location, while a wave is intrinsically spread out. This wave-particle duality is encapsulated in Heisenberg's Uncertainty Principle, according to which, while both states exist in a complete picture of reality, we can never focus on both simultaneously. For example, when the value for the position of a particle is measured to the greatest degree of accuracy, its momentum is not. Conversely, when the momentum is well defined, its position is not.

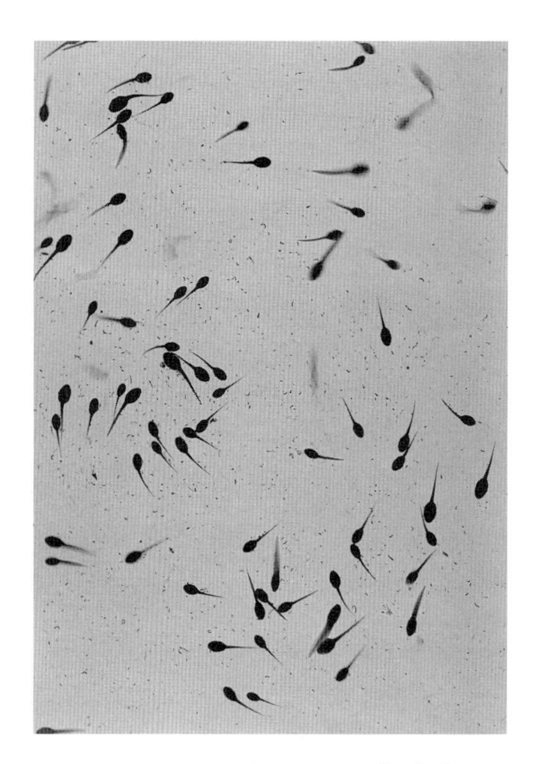

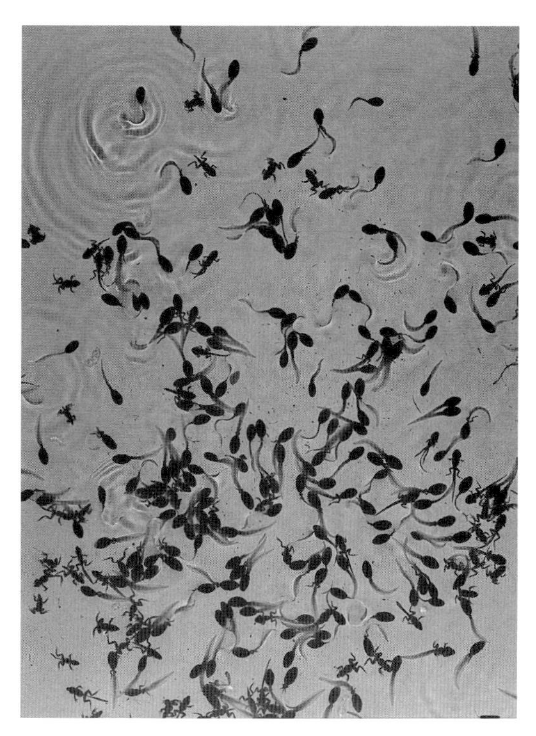

Susan Derges, details from *Spawn* (nos. 7 and 9), 1995. Series of 16 cibachrome photograms, 40.6 x 27.9 cm. © Susan Derges. Courtesy of the artist and Michael Hue-Williams Fine Art Limited, London.

Informed by these theories, two artists, working in entirely different media, have created artworks which exploit the nature of indeterminacy and, in doing so, they remind us that, even at the macro-level, nothing is fixed and there are always alternative versions of reality. In two series of photographic works entitled *Full Circle* (1992) and *Spawn* (1995), Susan Derges has recorded the process of metamorphosis from tadpole to frog. A shallow glass tank containing water and frog spawn was placed on top of a flat sheet of Ilfochrome paper, which produces positive rather than negative images. The tank was then exposed to light directed from above at regular intervals throughout the growth process. The images of the tadpoles and developing frogs were therefore recorded directly onto the paper below. From this very simple process, Derges showed how the resulting images depended on the time of the exposure. Sometimes the tank looked empty because the tadpoles or frogs were swimming too fast to be recorded but when a flashlight was used the tank appeared full. Each image provided a relative rather than absolute picture of what was going on. The choice the artist had to make was whether she should record the creatures' position or their momentum, making it clear that both were necessary for a full understanding of the process.

Antony Gormley has designed a sculpture, *Quantum Cloud,* to stand on four cast iron columns high above the Thames adjacent to the Millennium Dome in London. The sculpture represents a significant shift in Gormley's work to date, from a preoccupation with mass, volume and skin to a concern with air, energy and light. *Quantum Cloud* takes the form of a 30-metre-high elliptical structure made from a random matrix of identical elements, which condenses into a human body form that is visible from some angles and not from others. The structure is also highly responsive to atmospheric conditions. Its outer antennae vibrate and move in the wind, reinforcing the notion of flux and change. In this respect the work alludes to the transformation of classical physics to quantum reality, with emphasis being on the role of the observer as a fundamental participant, determining what is actually being seen.

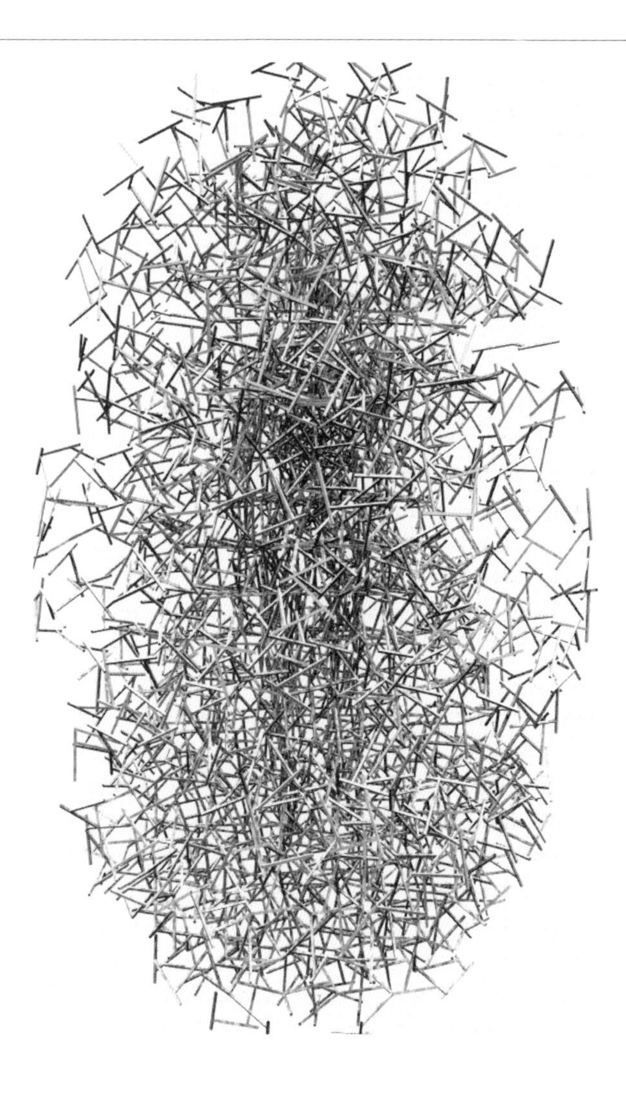

Antony Gormley, *Quantum Cloud,* work in progress. Random matrix of 3500 164-cm lengths of galvanised square steel tube, 30 m x 16 m x 10 m. Courtesy of the artist.

The role of the observer in quantum theory proved a great problem for Einstein. During the years spanning the late 1920s and the early 1930s he and Bohr fought a theoretical battle over the future of quantum physics. According to Bohr, one cannot make meaningful statements about a world divorced from observations and it is fruitless and impossible to speculate on the causal nature of phenomena on their own terms. One can speak only of one's observations or measurements. Indeed any refinement of quantum theory leads to wider, not narrower, applications: the greater the number of speculations the greater number of options are available to obtain a complete picture of reality. Quantum theory incorporates *indeterminism* as a fundamental feature. Einstein could not accept this outrageous randomness – 'God does not play dice', he famously said – and he regarded it merely as a statistical theory in that it dealt with the probable outcomes of similar experiments. Einstein wanted to believe in the possibility of a yet unknown fundamental and comprehensive theory which might provide missing information about the underlying structure of reality, a version which was ultimately separate from any observer's interventions. Even today the alternative theories coexist and although much refined subsequently, at the micro-level Bohr's 'orthodox theory of quantum mechanics' still prevails.

A theory of everything?

'I asked Einstein one day, "Do you believe that absolutely everything can be expressed scientifically?" "Yes," he replied, "it would be possible, but it would make no sense. It would be a description without meaning – as if you described a Beethoven symphony as a variation of wave pressure."'[19]

Unification has been the driving force in much of modern theoretical physics. It refers to the discovery that two phenomena which had previously seemed completely separate have a common origin. The first great unification in modern physics came in the middle of the nineteenth century, when the Scottish physicist James Clerk Maxwell discovered that electricity and magnetism were different manifestations of the same phenomenon, which he called *electromagnetism*. Furthermore, he discovered that the electromagnetic wave has a speed equal to the speed of light. Light is therefore taken to be one example of an electromagnetic wave. This was to have a profound effect on Einstein's development of his Special Theory of Relativity. Maxwell's discovery was the first step of a development in physics which led to an understanding that not only were electricity and magnetism different manifestations of a single phenomenon, but so too were the nuclear forces.

The now standard model of elementary particle physics began in the 1930s with the discovery that the several hundred different kinds of atomic nuclei were all composed of protons and neutrons. In the following two decades, as larger particle accelerators were built, hundreds more elementary particles were found. However, in the 1960s, it was proposed that most of these particles were not the most elementary manifestations of matter. Each is composed of even more fundamental particles which the physicist Gell-Mann called *quarks*. Quark theory formed yet another revolution in thinking which has lead to the theory of the strong nuclear force known as *quantum chromodynamics*.[20]

The idea of unification requires that different kinds of elementary particles are different manifestations of a small number of elementary entities, and string theory emerged in the mid-1980s as an explanation for this, with the great attraction that it also includes the gravitational force (the fourth of the four known fundamental forces of elementary particle physics). The basic premise of string theory is that the fundamental entities in nature are extremely small one-dimensional strings, whose different modes of vibration determine the different elementary particles. For consistency, string theory needs to include another type of unification known as *supersymmetry*, which combines the two distinct orders of elementary particles in nature: *fermions* (particles that make up matter, i.e. electrons, quarks, neutrons, protons) and *bosons* ('interaction' particles that are the carriers of the four fundamental forces, e.g. photons, W-mesons, gluons and gravitons).

The ensuing *superstring* theory is difficult for the lay-person to visualise – and the evidence for it is provided through mathematics. Superstrings are one-dimensional objects that exist in a ten-dimensional space-time (nine spatial dimensions, and one time dimension). Physicists find little difficulty in operating with so many dimensions in a mathematical way, but in order to accord with our daily experience, they are obliged to postulate that while nature does in fact have ten dimensions, the 'extra' six dimensions are 'curled-up' within the smallest quantum length possible (the so-called *Planck length*),[21] providing a tiny 'ball' of six-dimensional space associated with every point in our four-dimensional universe.[22] (To get some idea of the scale of these phenomena the reader should bear in mind that a string inside an atom has been compared to the size of a single atom inside the Solar System.)

Many different, mathematically consistent, string theories are known and it is not clear which, if any, is the physically correct one. However, there is an exciting conjecture that these are all different modes of a single underlying theory – known as 'M-theory' – which would provide a truly unified picture of all the forces of nature, as well as describing the origin of space and time themselves. At the time of writing, the search for this M-theory is one of the major research programmes in theoretical physics.

One of the few artists who has seriously addressed the physics of space, time and matter through artistic concepts and visualisations is John Latham. In work which is often difficult to categorise, Latham has deliberately overstepped the boundaries generally applied to art and makes references to physics, psychology, linguistics and anthropology. His artworks are not to be regarded as ends in themselves or art for art's sake but as devices for comprehending the universe and, crucially, the individual's place in it. In this respect, Latham regards himself as an 'incidental person' instead of an artist in the traditional sense of object-maker. For him, the artist is a person who 'stands at some distance from events and is capable of reflecting upon them critically, reaching insights by means of intuition'.

In 1988, Latham exhibited *History of Time* at the Lisson Gallery, London. In this work, Stephen Hawking's best-seller *A Brief History of Time* was sawn in half and mounted on both sides of a thick slab of glass placed on a wooden base. In dividing the book, Latham wished to symbolise 'the divided state conception which still informs scientific accounts of the universe'. The purpose of art, he believes, is to re-create what he regards as the lost relationship between the individual and the whole, as a result of a crisis in physics because of the limitations of theories 'which fail to include individual consciousness'.[23]

From the early works of the 1950s onward, Latham has formulated and developed an alternative theory of time which includes both the person and his motivations in *events*, which Latham considers to be the fundamental units of the universe, rather than particles. A distortion of language, a displacement of material objects and the inclusion of performance in his artworks have all been utilised in his ideas for a new artistic form of expression, as different from previous art practice as quantum theory is from classical physics. In a 1959 book entitled *The O-Structure*, Latham collaborated with animal behaviourist Anita Kohsen and astronomer Clive Gregory to initiate and develop a 'psycho-physical cosmology', which they derived (revised and updated) from the physicist Sir Arthur Eddington's mathematical model of quantum theory.[24] As in many of Latham's writings, the layout and language of scientific papers were appropriated, with formula-like expressions and diagrams to illustrate the text.[25]

In other work Latham assembled sawn-up, burned and painted books, together with metal fittings and machine fragments, which protruded from the picture surface. This created the impression that the objects were exploding out of the surface towards the viewer as a consequence of some kind of Big Bang. He subverted the normal temporal reading of books by presenting a randomised sequencing for the reader, so that the text could be read spontaneously and in no particular order. Latham referred to these assemblages as *Skoobs* – a form of literature to be read as what happens or happened, and to be looked at whole. As in so many of his works, in spray-paintings and *Skoob Films*, Latham is concerned with making present the trace of an event, where the change from object into 'action' is the result of a reflection on the role of the observer. Similar concerns governed Latham's public burning of book sculptures, in works called *Skoob Towers*, which provided a continuation of his book transformations. Experienced not as permanent art objects but as reversed 'event sculptures', they existed only at their moment of vanishing, making the viewer aware of the indivisibility of past, present and future.[26]

Latham's artworks form a conscious intellectual response to contemporary theoretical physics and he goes out of his way to set up encounters, public and private, with physicists so that he can address their theories from his resolutely individual perspective. Perhaps unsurprisingly, his intentions have often met with misunderstanding or incomprehension from the scientific community. At a public discussion with scientists at the Riverside Studios in February 1990, Latham confronted a panel of eight scientists, including biologists, physicists and cosmologists, in order to debate his ideas, which he preceded with a number of visual demonstrations. Although several of the panellists – somewhat patronisingly – praised Latham for tackling such a vast and difficult subject as space and time, they pointed out that some of his ideas about science were inaccurate, particularly since they could not be tested by experiment or proven through calculation – first requirements of the scientific method to which all scientists subscribe. Moreover, his theories did not propose testable hypotheses. As the event was held in an arts centre, the audience was drawn mainly from the arts community and therefore tended to side with Latham (when one scientist admitted he was baffled, an uproar went up from the audience). The only art critic on the panel, David Batchelor, ingeniously dubbed the event 'splatter physics'[27] and a sympathetic account subsequently provided by poet Karl Birjukov in *AND* magazine expressed dismay at the scientists 'lack of imagination'.[28] Perhaps inevitably, encounters like these between artists and scientists are bound to arrive at an impasse. Although Latham clearly has a sophisticated intellectual grasp of contemporary physics and pays deference to the scientific method, he does not 'believe' in it as a valid explanation for the way the universe is. As if to imply that science is merely another version, he describes his own theories as 'fiction', a term he holds in high regard. Art has as much validity as science when it comes to representing the 'truth'.

Latham is still regarded by both the scientific and arts communities as something of an eccentric. Nevertheless, scientists have sometimes been surprised at his intuitive grasp of complex systems, especially his ability to make visible concepts and events which are hard to visualise. In Latham's work *Time-Base Roller* (1972) he addressed the notion of time, viewing it as a 'construction line'. The work takes the form of a rotating cylinder over which striped and lettered canvas strips unroll as if to represent a continuing passage of time. The back of the rolled canvas represents the past from which elements can repeatedly be returned to the future. In operation the *Time-Base Roller* serves as a visualisation and demonstration of certain features of the physicist's notion of time. Latham has been involved in a number of meetings with theoretical physicists Chris Isham and Konstantina Savvidou at Imperial College, who were intrigued to find that Latham's ideas in *Time-Base Roller*, with its two concepts of time associated with the vertical and the horizontal axes, had a striking correspondence to their own diagrams of time as 'being' and

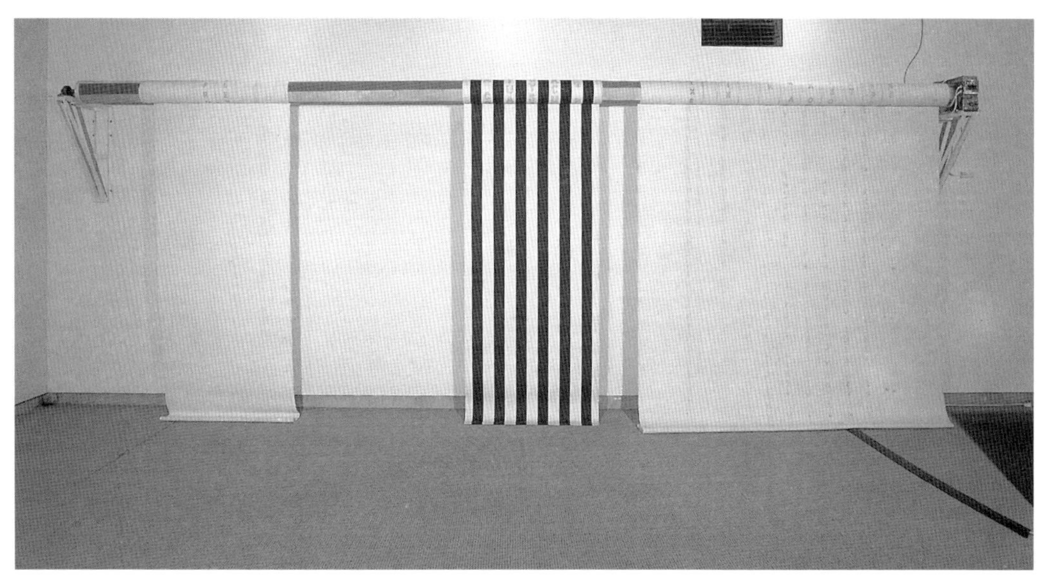

John Latham, *Time-Base Roller*, 1972. Canvas on motorised cylinder, 6 m, fall 428 cm. Courtesy of Lisson Gallery, London. Photo: Gareth Winters.

'becoming', which had arisen in their studies of the role of time in a new approach to quantum theory known as the 'consistent histories theory'.

Scientists often use visualisations to order to make their thought experiments more graspable but Latham undoubtedly possesses an exceptionally intuitive understanding of some of the theoretical processes.

From the simple to the complex

'The central aim of science is to render the complexities of the universe transparent, so that we can see through them to the simplicities beneath ... But at the bottom of the funnels that lead downward from the complexity of our daily lives, we find two apparently contradictory things: order and chaos.'[29]

Over the past 15 to 20 years researchers from many scientific disciplines have realised that there are limits to our ability to predict events in nature. Although the universe can be described simply in terms of natural laws, it exhibits a disposition for disorder, complexity and unpredictability. 'Chaotic' behaviour has been observed in a wide variety of natural systems, from beehives to weather patterns. However, despite its name, Chaos Theory seeks to eliminate rather than discover or create chaos and it analyses processes which appear chaotic on the surface but which, on detailed examination, prove to manifest subtle strands of order. Studies of chaotic systems have revealed that tiny fluctuations from an apparently stable state begin gradually to increase and then exhibit more and more changes, resulting eventually in totally new organisations of structure. These self-organising systems, as described by chemist Ilya Prigogine, exist in *far from equilibrium conditions*, delicately poised between order and chaos, and it is this poise which stimulates the impulse to create order once more.[30] This is another example of how reality can be seen not as a fixed entity but as a process and a creative evolution. Within this evolutionary paradigm, the concept of time is of central significance. For humans, the idea that time flows forward is implicit in reality, as in 'time's arrow'. All activity occupies time but not all of it flows consistently forward. Some systems are apparently cyclical or vacillate back and forth between chaos and order, although none ever returns to exactly the same state that it was in before.

The British artist Andy Goldsworthy creates artworks by working with nature *in situ*. At the heart of his working method is the process of looking, touching, trying things out, and discovering the hidden potential for change. He writes:

> For me, looking, touching, material, place and form are all inseparable from the resulting work. It is difficult to say where one stops and another begins. Place is found by walking, direction determined by weather and season. I take the opportunities each day offers.[31]

Stones
dipped in water
then frozen to quarry face
made slowly over three cold days
place untouched by the sun in winter

Andy Goldsworthy, *Stones*, Pickering, Yorkshire, 23–25 December 1992. Courtesy of the artist.

> Change is the key to understanding. At its most successful, my 'touch' looks into the heart of nature; most days I don't even get close. These things are all part of a transient process that I cannot understand unless my touch is also transient – only in this way can the cycle remain unbroken and the process complete. Movement, change, light, growth and decay are the lifeblood of nature, the energies that I try to tap through my work.[32]

An awareness of time plays an integral part in all of Goldsworthy's works and it is exploited at many different levels. The natural materials he works with are moulded by time as is his making of the work. By capturing the 'event' in mid-process, whether at any stage during the making of the actual work or fixed at a single point through photographs, the artist highlights the work's impermanence and implies that all the materials used, whether they are stones, leaves, sand, or sticks, will eventually fall back to the earth. A sense of randomness and unpredictability is implicit even in the more formal works where there is a rearrangement of nature into geometrical forms – using stones, leaves or flowers of a certain colour. We know that the fall of leaves or the distributions of stones on a beach are random. Colour in nature is never static. As seasons change, as things die, so colours change or drain away. The colours found and the order in which they appear can never be predicted, but depend on their location, the time of year and the vagaries of climatic change, never the same from one year to the next, one season to the next or one moment to the next. All Goldsworthy's works, like their natural subjects, exist in a tension between imbalance and collapse, hovering with built-up energy. There is a delicate balance between stability and instability, between order and chaos, and this balance is never static, but dynamic. Although Goldsworthy does not consciously address the science of chaos theory, his work demonstrates its processes more vividly than could any rational explanation.

Problems in communication?

'There is a problem in admitting to someone that you are a physicist, apart from either being tarnished with the ghost of Oppenheimer or having provoked a look of boredom, it's that you are constantly aware of the limitations of words.'[33]

The scientific mind exhibits an insatiable curiosity and the desire to impart understanding. Science is rarely easy to communicate, however. A proper understanding of scientific concepts may take years to acquire and certain terms assume a wealth of background knowledge. Often ideas are expressed exclusively through mathematics. Even so, it is striking to note how often visual images are used to communicate scientific principles. Indeed, a large part of scientific expression involves representations which are either wholly or partly non-linguistic in character, such as charts, graphs, diagrams and physical models. The immediacy and economy of these modes of representation is self-evident. It is possible to provide a verbal description of an electric circuit, for example, but it would be so long and complex as to make it almost unintelligible. A diagram is much clearer. Scientists use images in specific ways to communicate what they have learned. However, understanding the content of a scientific illustration requires some intellectual knowledge of the phenomenon to be studied, as well as an understanding of its labels and terms.

Humans are highly visual animals and we often think in pictures. From cave paintings to the computer, the visual image has assisted the human race in describing, classifying, ordering, analysing and ultimately reaching a greater understanding of the world. Images trigger an internal response, where the viewer transforms the static image into an intellectual or an emotional experience. Our ability to read visual images demonstrates our power to think in the abstract. Such visualisations give rise to the metaphors which fix them in language. Scientists sometimes invent new names for newly discovered phenomena – 'quarks', 'bosons', 'fermions' – but they also rely on analogues which make implicit a comparison with a familiar thing. In a commonplace way we can almost envisage superstrings, black holes, the speed of light or the boundary of the universe. However such familiarity can be misleading, for in physics words should not be taken too literally. Some terms also have different meanings for different scientific disciplines. *Adaptation* has a different meaning for the biologist and the psychologist, as does the term *environment* for the physicist and the botanist. *Chaos* does not mean randomness, and the *Uncertainty Principle* does not mean everything is uncertain. *Fuzzy Logic* or *Relativity* do not mean that 'anything goes'. In scientific terms, *dark matter* refers to the hypothetical matter which, according to current theories of cosmology, makes up 90 to 99 per cent of the mass of the universe, but so far remains undetected.

There are deep and genuine connections being made between artists and scientists, but sadly, there are also some which rely on clumsy analogues or which have degenerated into empty metaphors. For the most part, any link between physics and the contemporary art world appears to be a one-way process, from physics to art, with artists borrowing concepts and words from physics to meet their own artistic endeavours. Schrödinger's Cat seems to be a particular favourite. In an installation of the same name, the artist Ansuman Biswas lived in a closed box for a week (and came out, not unexpectedly, alive).[34] On the same theme the theatre company Reckless Sleepers provided a blank space where 'anything is conceivable', but where, presumably, nothing visibly unpredictable occurred.[35] In a 1998 exhibition, *Dark Matter: A Visual Exploration of New Physics*, at the Harris Museum and Art Gallery, Preston, the intention was to show 'how artists have taken current issues in science as the starting point for their own work'.[36] Artworks were entitled *The Wormhole* (Michael

Petry), *Virtual Absolute: Cosmos Series* (Stephen Hughes) and *Galaxy Classifications* (Russell Crotty), for example, but at least the participants recognised that the physics was a starting point only and one should not expect too literal an engagement. It would be an exceptional artwork which seriously took on the mathematical dimensions of physics, for example. Equations with names such as the Jones Polynominal, the Codazzi-Raychaudhuri identity, de Raham cohomology classes, Betti numbers and the Cheeger-Gromoll Splitting Theorem are a closed book to most of us, even while they sound intriguing.

It is interesting to ponder quite how far any literal engagement with science can go or whether it would be useful or enlightening for practitioners on either side. What is certainly the case is that some artists, whether consciously or not, do seem to be finely tuned to conceptual changes in science. Indeed critical theorists would claim that there are cultural and philosophical reasons underpinning the coincidence in imagery in both science and art. When there are changes to cultural and political perspectives, it may be no surprise that ways of analysing and representing the world reflect these new cultural stances. The commonly accepted interpretation for the development of perspective in art, for example, is that it is a manifestation of Renaissance individualism, placing Man at the centre of his world. This coincided with an increasing confidence in Man's role in understanding, interpreting and gaining control of nature, as manifested in early science, or 'natural philosophy'. Cubism, by contrast, developed during a period of political breakdown and unrest, and portrays the world as a precarious interplay of independent units, each coherent and lawful only unto itself and where the viewer is as much participant as observer. Some historians of science have remarked on an affinity between abstract thought in art and abstract form in physics, particularly at the beginning of the twentieth century when radical theoretical changes were occurring in each discipline.[37] However there is little evidence to prove that much direct communication between early twentieth-century artists and scientists existed and it is more likely that any associations – for example, between Cubism and relativity theory – were simply the result of a cultural coincidence,[38] another example of what art historian Martin Kemp calls the artist's 'structural intuitions'.[39]

If there is to be any dialogue between art and physics it is important to realise that they show fundamental differences in approach and that a great deal in the study of each is non-transferable. The quest for simplicity on the part of science and the delight in complexity on the part of art are incompatible although each side can learn from the other. The complex nature of reality goes far beyond the boundaries of any single language, and incorporates both the simple and the complex. It is a view articulated by cosmologist John D. Barrow:

> The study of human actions, human minds, and human creativity has been quick to see complexity; slow to appreciate simplicity. Science, quick to see uniformity, has at last begun to appreciate diversity; but there is

> much for the creative arts to learn from the unity of the Universe about the propensities of your senses and the sights and sounds that excite them. And science, in its turn, will discover much about the emergence of complex organised structures from a renewed study of the mind's most artful inventions. This is a place where two ways meet.[40]

In the early 1960s Richard Feynman had many meetings with the artist Jiiayr Zorthian to discuss (and often argue about) the relationship between the arts and sciences. Zorthian would attack science by claiming that scientists destroy the beauty of nature by 'picking it apart and turning it into mathematical equations'. Feynman replied that, far from destroying nature's beauty, scientists actually increased our pleasure in it. He stated:

> There is an aspect of generality that you feel when you think about how things that appear so different and behave so differently are all run 'behind the scenes' by the same organisation, the same physical laws. It's an appreciation of the mathematical beauty of nature, of how she works inside; a realisation that the phenomena we see result from the complexity of the inner workings between atoms; a feeling of how dramatic and wonderful it is. It's a feeling of awe.[41]

Rowan leaves laid around hole
collecting the last few leaves
nearly finished
dog ran into hole
started again
made in the shade on a windy, sunny day

Andy Goldsworthy, *Sunflower*, Yorkshire Sculpture Park, West Bretton, 25 October 1987. Courtesy of the artist.

Notes and References

1 Jeremy Hooker, from 'Workpoints', published in exhibition catalogue *Groundwork: Sculpture by Lee Grandjean and poems by Jeremy Hooker* (University of Nottingham Arts Centre, Djonogly Art Gallery, 1998).

2 R. Feynman, R. Leighton and M. Sands, eds., *The Feynman Lectures on Physics*, vol. 1 (Reading, MA, Addison-Wesley, 1970).

3 John D. Barrow, *Pi in the Sky: Counting, thinking and being* (Oxford, Clarendon Press, 1992).

4 John D. Barrow, from Interalia lecture held at Marks and Spencer, Michael House, Baker Street, London, 1996.

5 Barrow (1992), p. 159, see note 3.

6 *Cameraless Photography: Susan Derges and Garry Fabian Miller*, exhibition catalogue (Japan, Yokohama Museum of Art, 1994).

7 Professor John Wheeler, 'Law without Law', from *Structure in Science and Art*, ed. Peter Medawar and Julian Shelley. Proceedings of the Third C.H. Boehringer Sohn Symposium held at Kronberg, Germany, 1979 (Amsterdam, Excerpta Medica, 1980), pp. 132–168.

8 Newton's Three Laws of Motion are the basis of all mechanics. In summary, they state that:
a. If no force acts on an object, or if all the forces are balanced, the object will stay in the same position, or carry on moving at a constant speed in a straight line.
b. If an unbalanced force acts on an object it will accelerate in the direction of the force according to the relation: Force = mass multiplied by acceleration ($F = ma$).
c. Whenever a force acts on an object an equal and opposite force acts on the agent producing the original force.

9 Rudolf Arnheim, 'Space', in *Art and Visual Perception: A psychology of the creative eye* (Berkeley and Los Angeles, University of California Press, 1954, 1974), pp. 218–302.

10 The Michelson-Morley experiment (1882) was designed to determine the speed of the earth relative to the hypothetical *ether* (a medium that, according to classical physics at the time, was assumed to permeate space and fill the spaces between particles and matter). The experiment produced a null result. Einstein's Special Theory of Relativity replaces the concept of *ether* by the postulate that the speed of light is a universal constant in every *inertial frame of reference.*

11 David Bohm, *The Special Theory of Relativity* (London, Routledge, 1996), and Max Born, *Einstein's Theory of Relativity* (New York, Dover Publications, 1965).

12 Bernard Riemann (1826–1866) gave convincing evidence, through geometry and mathematical equations, that the structure of space in the physical world is determined by the distribution of matter.

13 Rudolf v.B. Rucker, *Geometry, Relativity and the Fourth Dimension* (New York, Dover Publications, 1977).

14 A. Lorentz, A. Einstein, H. Minkowski and H. Weyl, *The Principle of Relativity* (New York, Dover Publications, 1952).

15 Richard Feynman, *The Character of Physical Law* (London, BBC Publications, 1965), p. 129.

16 Lee Smolin, *The Life of the Cosmos* (London, Weidenfeld and Nicolson, 1997), p. 64.

17 Roy Frieden, *Physics from Fisher Information* (Cambridge, Cambridge University Press, 1999), p. 24.

18 Walter Moore, *A Life of Erwin Schrödinger* (Cambridge, Cambridge University Press, 1994), pp. 219–20.

19 Conversation between Frau Born (widow of physicist Max Born) and Albert Einstein, recounted in Ronald W. Clark, *Einstein: The Life and Times* (London, Hodder and Stoughton, 1973), p. 191.

20 Quantum chromodynamics (QCD), a theory describing the interactions of quarks, the elementary particles that make up all hadrons (i.e. subatomic particles such as protons and neutrons). In quantum chromodynamics, quarks are considered to interact by exchanging particles called gluons, which carry the strong nucleur force, and whose role is to 'glue' the quarks together. The prefix 'chromo' refers to 'colour', a mathematical property assigned to quarks. Another facet of unification was the discovery in the 1970s that the electromagnetic force and the weak nuclear force are different manifestations of one underlying force. The theory of quantum chromodynamics is compatible with this structure, but so far attempts to unify all three forces (the electromagnetic, the weak nuclear force, and the strong nuclear force) have not been successful.

21 Planck length: the length scale in which the quantized nature of gravity should first become evident, i.e. approximately 10^{-33} cm.

22 Paul Davies and Julian Brown, eds., *Superstrings: A theory of everything?* (Cambridge, Cambridge University Press, 1992).

23 John A. Walker, *John Latham: The incidental person – his art and ideas* (London, Middlesex University Press, 1995).

24 Sir Arthur Stanley Eddington (1882–1944), British astronomer and physicist, who did important work in relativity and astronomy. Eddington helped to clarify the theory of relativity, and made mathematical contributions to the subject. He was best known as a populariser of science, and his work *The Nature of the Physical World* (1928) was one of the most widely read books ever on abstract science.

25 C. Gregory and A. Kohsen, *The O-Structure:*

An introduction to psychophysical cosmology (Church Crookham, Hants, Institute for the Study of Mental Images, 1959).

26 John Latham, *Publishment of Skoob, supplement to John Latham: Assemblages,* exhibition catalogue (Oxford, Bear Lane Gallery, 1963).

27 D. Batchelor, 'Splatter Physics', in *Artscribe International*, 82, Summer 1990, pp. 7–8.

28 K. Birjukov, 'Time Versus Space', in *AND: Journal of art and art education*, 22, [illegible], pp. 16–17.

29 Jack Cohen and Ian Stewart, *The Collapse of Chaos: Discovering simplicity in a complex world* (London, Penguin Books, 1995), p. 29.

30 Ilya Prigogine and Isabelle Stengers, *Order Out of Chaos: Man's new dialogue with nature* (London, Heinemann, 1984), p. 176.

31 *Andy Goldsworthy* (London, Viking Penguin, 1990).

32 Andy Goldsworthy, from a talk given at Interalia Conference, 'The Nature of Colour – East and West', Arnolfini, Bristol, 1991.

33 Basil Hiley, Professor of Theoretical Physics, Birkbeck College, University of London, in conversation with Richard Bright. Robert J. Oppenheimer (1904–1967) was an American physicist and government adviser, who directed the development of the first atomic bombs at Los Alamos, New Mexico, 1943–5.

34 Performance by Ansuman Biswas entitled *CAT*, at the South London Gallery, 1998, organised by the Arts Catalyst, London. Ansuman Biswas performed an experiment/demonstration inspired in part by Schrödinger's Cat, the famous paradox in quantum physics. The work arose from a comparative study of modern scientific methodology and the 2,500 year old Indian science of *vipassana* (according to Mahayana Buddhism, an analytical examination of the nature of things that leads to insight into the true nature of the world). It lasted for ten days during which the artist attempted to maintain close, continuous observation of all physical phenomena.

35 Performance by Reckless Sleepers at Arnolfini, Bristol, 11–12 February 1999. Reckless Sleepers constructed a chamber in the shape of a box, and named it in honour of quantum physicist, Erwin Schrödinger. They described the box as 'an experimental chamber, a crucible, a television, a tuning device... a place to concoct dreams for insomniacs from captured thoughts, and where yes means no'.

36 From the foreward to *Dark Matter: A visual exploration of new physics*, exhibition catalogue (Preston, Harris Museum and Art Gallery, 1998).

37 Arthur I. Miller, *Insights of Genius: Imagery and creativity in science and art* (New York, Springer-Verlag, 1996).

38 Linda D. Henderson, *The Fourth Dimension, and Non-Euclidean Geometry in Modern Art* (Princeton, New Jersey, Princeton University Press, 1983).

39 '... the term "structural intuitions" is to be understood as carrying a two-fold reference: to the way in which natural forms and phenomena exhibit distinct and perceptible patterns, often shared across inorganic and organic phenomena over the widest possible range of scales; and to the way that we have been endowed with processes of mental sifting and structuring that enable us to generate patterns on geometrical bases and logical sets of numbers.' Martin Kemp, 'The Music of Waves. The Poetry of Particles', in *Susan Derges, Liquid Form 1985–99* (London, Michael Hue-Williams, 1999), pp. 8–9.

40 John D. Barrow, *The Artful Universe* (London, Penguin Books, 1997), p. 246.

41 Jagdish Mehra, *The Beat of a Different Drum: The life and science of Richard Feynman* (Oxford, Clarendon Press, 1996), p. 580.

8

Inside – Outside – Permutation: Science and the Body in Contemporary Art

Andrea Duncan

Now, in the firm hands that quiver with a careful strength,
Your knife feels through the heart's transparent skin; at first,
Inside the pericardium, slit down half its length,
The heart, black-veined, swells like a fruit about to burst,

But goes on beating, love's poignant image bleeding at the dart
Of a more grievous passion, as a bird, dreaming of flight, sleeps on
Within its leafy cage – 'It generally upsets the heart
A bit, though not unduly, when I make the first injection.'

JAMES KIRKUP, from 'A Correct Compassion – To Mr Philip Allison, after watching him perform a mitral stenosis valvulotomy in the General Infirmary at Leeds'.[1]

Contemporary science has transformed our view of the body as never before, dislocating and fragmenting our sense of personal identity, of boundary, scale and time. As scientists engineer and reconfigure the body at improbably minute and delicate levels, there lurk in our minds vestigial traces of the alchemist's homunculus and Mary Shelley's monster. Like artists, scientists transmute and re-create. Like scientists, artists have begun to appropriate a quasi-scientific neutrality and detachment, playing with investigative instruments, precision technology and deep computing and, in so doing, questioning and unsettling the me/not me body relationship. Some of the most difficult subject matter has been transmuted through photography, video and film and, as spectators, we stand on the hygienic side, sufficiently distanced to look into another's body processes and see – by the very act of framing – real bodies objectified as works of art. Yet the most profound work slips from a neutral dialogue with the scientific into the metaphysical, provoking fundamental questions about the nature of identity, and ultimately, of *being*.

The body first became a specific site of avant-garde art in the early 1960s and continued to develop as an art form, mainly through performance, throughout the 1970s. While artists such as Karen Finley, Marina Abramovic, Joseph Beuys, Bruce Nauman, Chris Burden and the Vienna Actionists were largely reacting to the existential malaise of Western post-industrial society rather than becoming excited by its technological innovations and discoveries, none of this earlier activity should be forgotten when we look at current scientifically grounded work with the body. If the art being made now has a cooler edge – clinical and distanced – this is because it presents the ideas of artists who have been informed and influenced not only by scientific research but, increasingly, by scientific method and its technology. Footage of the French choreographer Kitsou Dubois 'flying' in the zero gravity of parabolic flight presents an example of an artist utilising a new environment in a working relationship with the French astronaut training programme, while the artist Orlan, as will be seen, has appropriated the operating theatre and expertise of plastic surgeons – relationships hard to imagine during the 1960s. Increasingly, artists recognise that if they are to question the parameters of body/identity/mind it is to science they must turn, not necessarily for a solution but for the most pertinent framing of the questions.

This is my body, and this is my software[2]

It may have taken the wolf 30,000 years to become a Pekinese dog but it has taken the French artist Orlan only ten years to metamorphose herself from her own 'raw' body material. During the 1960s Orlan developed a style of performance art which had been pioneered by some American feminist contemporaries with a focus on the female body as a means of expression. By the end of that decade, however, Orlan had arrived at a more independent position, presenting quasi-religious aspects of feminine *jouissance*[3] through her self-baptism as *St Orlan* and a first sculpting of herself in the traditional artist's material, marble (1972). In 1978 there was a paradigmatic shift when the performance work became re-contextualised through the contemporary technologies of surgery and film. During an emergency operation for an extra-uterine pregnancy, Orlan used video to re-present the surgical action as performance.[4] This surgical/technological approach increased in importance for her throughout the 1980s, and in 1990 the surgical operations began which were to transform her physical appearance. Her work *This is my body, and this is my software* (1990) is perhaps the most eloquent and distinctive reflection on a secular age of physical mutability *sans* religion. Orlan's operations are visually layered in metaphor and have included the shamanic/carnivalesque theatrical display of earlier work – now, however, transferred to the sterile and cool 'theatre' of medical surgery. In a protracted series of operations, the surgery has included liposuction and extensive alterations to her face. While, as she admits, 'the operating act is outside amusement', it would nevertheless contribute to 'a performance orientated towards the future, a radical performance for myself and beyond myself'.[5] The operations are a way for her to work with ideas which are already 'incarnated' in the flesh. Fundamentally, these surgical interventions 'ask questions about the status of the body in our society and its future in coming generations in terms of new technologies and the genetic manipulation which will not be long in coming'.[6]

Orlan has therefore brought a timely focus to bear on the pseudo-religious and alchemic nuances of the laboratory/operating theatre. These places have become, as neuroscientist Steven Rose describes elsewhere, the 'ideological and technological powerhouses of modern society',[7] yet they retain something of the residue of the priest's sanctum and of the precious tools and procedures of ritual. Each of Orlan's operations, videoed as a performance piece, includes the artist, usually under local anaesthetic, giving instructions and interpretative direction. This has included her own readings from Kristeva's *Powers of Horror* and the texts of Antonin Artaud,[8] while she and her entourage have been dressed in the haute couture of Paco Rabanne and Issey Miyake. Through the medium of video these images have been relayed live to audiences and the resulting surgical detritus (fat, blood and flesh) has been treated as both surgical specimen and as reliquary, in a quasi-ritualistic sense. Post-operative photographs which clearly show bruising and stitches are part of the documentation. These, while replicating illustrated medical texts, also recall art historical images of religious martyrdom and imply the spiritual transformation of the alchemist's base matter.

The sixteenth-century German alchemist Paracelsus devised very careful recipes for the creation of the homunculus, including the combinations of urine, semen and blood leading to their transformation in the hermetic vessel. The homunculus – that first cloned being which represented the alchemist's longing for transcendence – has been revived in the art of Orlan's performance and she has brought to the cool and insensitive world of surgical theatre something of the hysterical inchoate world of rude being. In that respect, while

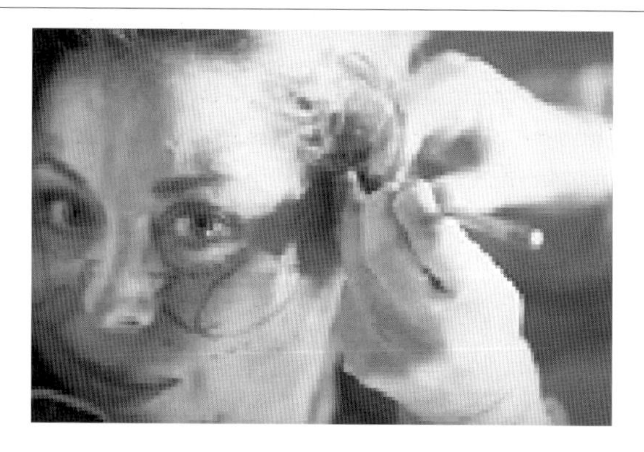

Orlan, Still from *Omnipresence* 1, 1990. Courtesy of the artist.

ordering and controlling the science of her face, she has let in the unmentionable and unspeakable and revealed science's ignorance of the metaphors unintentionally implicated in its technology. Referring to her operations, she argues that she wanted 'to intervene in the cold and stereotyped image of plastic surgery, to alter it with other forms'.[9] At the worst moments of Orlan's facial transformations, we see the mask of the face being lifted away from the underlying and formless mess of blood, tissue and bone. This invokes a primitive dread which even the clean-drawn incisions of the surgeon's scalpel cannot excise. At the moment when Orlan's face lifts away from her ear, as Parveen Adams observes in her essay 'Operation Orlan', 'something flies off, this something is the security of the relation between the inside and the outside. It ceases to exist.'[10] Yet such is the awesome efficacy of technology that, having undergone such terrible rendering and cutting, Orlan appears again made whole. The gulf remains visibly unbridgeable, however, in her conference presentations of the surgery, where, as Adams suggests, 'the contrast between this Orlan, theatrical and artificial in her long gown, in possession of herself, and the other Orlan cut, bleeding and supine on the operating table, is stark.'[11] Yet the contrast is already inculcated in the course of the operation, with the voice and its readings from the chosen texts emerging from the body's tissue and from the midst of the surgical intervention. Orlan has usurped the surgeon's place, for that disparity between image and presenter, between the abject, incapacitated patient and the medical practitioner is undone by the conscious and controlling voice which is Orlan's most prophetic moment in the attempted transcendence of the flesh: 'During surgery I read texts as long as possible, even while my face is being operated on. In the most recent operations this produced an image of a cadaver under autopsy which keeps on speaking, as if the words were detached from its body.'[12]

What does it mean to become part of the 'theatre' of the laboratory?[13]

The work of the Canadian artist Louise Wilson offers some interesting parallels to that of Orlan but Wilson focuses more on the experience of offering herself as a subject in another's medical research. Here, access to the 'privileged technological spaces' becomes possible and she has volunteered herself for research into sleep, dream and memory, and high-tech investigative techniques such as MRS (magnetic resonance spectroscopy) and MRI (magnetic resonance imaging) in which, as she points out, 'the body is essentially acted upon and rendered as physics. MRI is an intervention which occurs below the level of the cells.'[14] One of the issues she raises is the disturbing experience of separation between consciousness and the body, the sense of oneself as simultaneously both conscious subject and inanimate object, both sensate body and mere data.

> To be the subject in a lab study is potentially to see one's body mediated as a generic site: one where physiological (and psychological) functions can be both disclosed and effected by invasive and non-invasive technologies alike... In the laboratory, the lived body, along with its virtual offspring, the data body, separate.[15]

Wilson's investigation into the experience of the research subject led to the installation *Possessed* (1995), where a staged laboratory-cum-psychoanalyst's consulting room invites the participant to recline on a couch while placing him or her just beneath a television monitor. The screen is close to the face and the accompanying audiotape offers an incantatory and seductive monologue. The 'patient' or viewer hears a recording on an alternating loop of the voices of two hypnotists – a French woman and an English man – and their relaxation techniques are followed by a specific test: 'The viewer is then asked to "anaesthetise" [his or] her right hand and simultaneously "erase" the target: a pink "ganglion" visible in the cut-away structure of a 3-D brain. After about four minutes of repeated suggestion, the "target", a visual representation of pain, is almost completely erased.'[16] This test was taken from PET/MRI scan work at the McConnell Brain Imaging Centre, Montreal Neurological Institute, and Wilson's concern in this staging was that, like the research subject, 'the subject of *Possessed* is enmeshed in and manifests the logic of the experimenter's study'.[17] Here, the subject, while still being affected by the piece, produces a 'feedback loop' in which he or she and the piece affect each other. Just as the test does not simply monitor the subject but has in itself a definite physiological and psychological effect, so it, in turn, is fundamentally affected by alterations which result from the patient's psychological response and 'the mapping of an activity is registered in the map itself'.[18] To emphasise this, Wilson points to recent research in the neurosciences which suggests that extended use of new image technology affects the subject and produces in the brain 'physical re-mappings of neural connections'.[19] In essence, the subject's physical brain is altered by extended exposure to neurological experiment. In this, the influence of the environment upon the subject, as Wilson demonstrates, takes on the eerie ability literally to shape the interior forms of mental personality. Wilson also highlights the anxieties of the 'digital physician' who, wholly removed from tactile contact, becomes a disembodied mind and eye, dealing with the largely symbolic presentation of another body. She quotes Michael Heim's observation that 'a psycho-technological gap opens up between doctor and patient. Surgeons complain of losing hands-on contact as the patient evaporates into a phantom of bits and bytes.'[20]

Other artists who have questioned the mind/body relationship in terms of human identity include Helen Chadwick, Damien Hirst, Marc Quinn and Mona Hatoum. In the last decade, the head, like the body, has become 'medicalised' and philosophical questions (if such they are) have frequently been posed via a scientific objectification of the artist's head. In Chadwick's *Self-Portrait* (1991) the human brain is shown exposed but remains held gently in the artist's hands. Hirst's *Dead Head* (1991) is a black-and-white photograph of the artist's laughing head juxtaposed with the quiet reflective smile of a severed (male) head which has obviously been the subject of dissection. As Hirst himself has remarked: 'You worry about your complexion but you'll be a skull in less than a hundred years.'[21] Marc Quinn's *Self* (1991), like so much in current medical research, requires refrigeration equipment to keep it viable. Formed from a series of extractions of Quinn's own blood, the head, made from a cast of Quinn's own, is placed behind perspex and presented on a base like any marble bust (see page 161). The base, however, is a refrigeration unit and the spectator is reminded of the chilled states of suspended animation which are routine in contemporary biomedicine – the harvested eggs, cells, and organs upon which so much human happiness now depends. Quinn has taken the concept of the self-portrait into the biotechnical age with the use of his own blood which, of course, offers us his blood type and genetic fingerprint – in itself a huge field of future research for the artist. Mona Hatoum's *Socle du Monde* (1992–3) is a large cube on the surface of which are undulating patterns made by iron filings influenced by magnets, producing a cold but strangely animated and rippling, furrowed surface. Visually like the cortex of the brain, the activity of the filings suggests the ruminations of a majestic intelligence, yet these undulations are also typical of peristaltic waves wholly absorbed in an intestinal and primary digestive process, far below the level of consciousness.

Of course, I *was* invading people's boundaries[22]

Mona Hatoum has been both attracted and repelled by medical intervention and she exploits the techniques of invasive science to demonstrate what she regards as her own 'penetrating gaze'. For her work *Corps étranger* (Foreign Body; 1994), she submitted the most intimate parts of her body to endoscopy and coloscopy for an installation piece in which she presents a video of these internal investigations projected on to the floor of a cylindrical viewing chamber, accompanied by a soundtrack of heartbeat and breathing. 'I wanted to give the feeling that the body becomes vulnerable in the face of the scientific eye, probing it, invading its boundaries, objectifying it', she writes. 'I felt that introducing the camera, which is a foreign body, inside the body would be the ultimate violation of a human being, not leaving a single corner unprobed.'[23] The endoscopic invasion is a slippery journey into soft and impressionable matter which emphasises the loss of physical integrity and in which, as Hatoum suggests, one glimpses 'the edge of an abyss that can swallow you up, the devouring womb, the vagina dentata, castration anxiety...'[24]

While the viewing booth for *Corps étranger* is partly a device for presentation, it is also reminiscent of a number of technological 'pods' – the automatic public lavatory, the laboratory cubicle and the space capsule. These echoes of other machines include the 'projectile eye' hinted at in Hatoum's preliminary drawing, and the final presentation in the booth recalls her 'scientific eye' and the essential curiosity which she articulates and the machine itself encapsulates, with its anthropomorphic extensions in subtle, facilitating tools. We can explore this further with Hatoum, when she recollects, in an interview with the critic Claudia Spinelli, that 'as a child, with a pair of binoculars, I used to spy on people passing by in the street. I used to wonder what would happen if my binoculars could see through the clothes and through the skin and through the flesh and through the bones.'[25]

In *Corps étranger* the invasive movement of endoscopy replicates the ceaseless drive of living organisms and, beyond that, the eternal flux and flow of systems at the borders of the animate.[26] This work with its links to medical procedure should also, however, remind us of the preoccupation with the visible forms of physical well-being in the Western world now that the routine threat of infectious disease and sudden and fatal illness has diminished. Yet despite advances in investigative technology, the public is still mystified by complex physiological processes. In this respect the content of the artwork *Corps étranger* has a resonance with Peter Greenaway's fictionalisation of the same procedure in his 1987 film *The Belly of an Architect.* Here a hospital examination of the architect Kracklite's developing stomach cancer is counterbalanced by his own ignorance of physiology. When informed that human intestines are 27 feet long Kracklite attempts to visualise this with an appropriate length of rubber tubing gathered up and pressed against his stomach. The random disorder of this tubing makes Kracklite observe, 'I used to marvel at … the human body … not any more… The intestines seemed pushed into the stomach cavity anyhow.'[27] This unpleasant comparison is, of course, made in relation to the solid and elegantly ordered world of architecture which Kracklite knows well.

In her work *Entrails Carpet* (1995), Hatoum focuses on the 'meandering shapes that looked like entrails' in the earlier *Socle du Monde* (1992–3).[28] For this, she has created an oversize Petri dish large enough for the human subject but crammed with intestines. These are, however, made of delicate pearl rubber and transcend the original, appearing as clean and antiseptic as any clinical environment. At another level, however, the piece suggests a domestic interior and the entrails disguise themselves as the pile of a pearl-pale bedroom Axminster, suggestively soft and squidgy to the feet, balancing the scientific references and the objective and isolating practices of the laboratory with the homely/unhomely nature of the domestic interior. In *Marrow* (1996), echoes of early childhood are evident in the baby's cot, but the traditional bars are replaced by a pale, floppy, rubbery material suggestive of bone marrow and the pliability of a flesh hardly formed. Skeletal and empty before, the cot no longer stands but is now fallen – and corporeal – the bars themselves bear the vestigial traces of

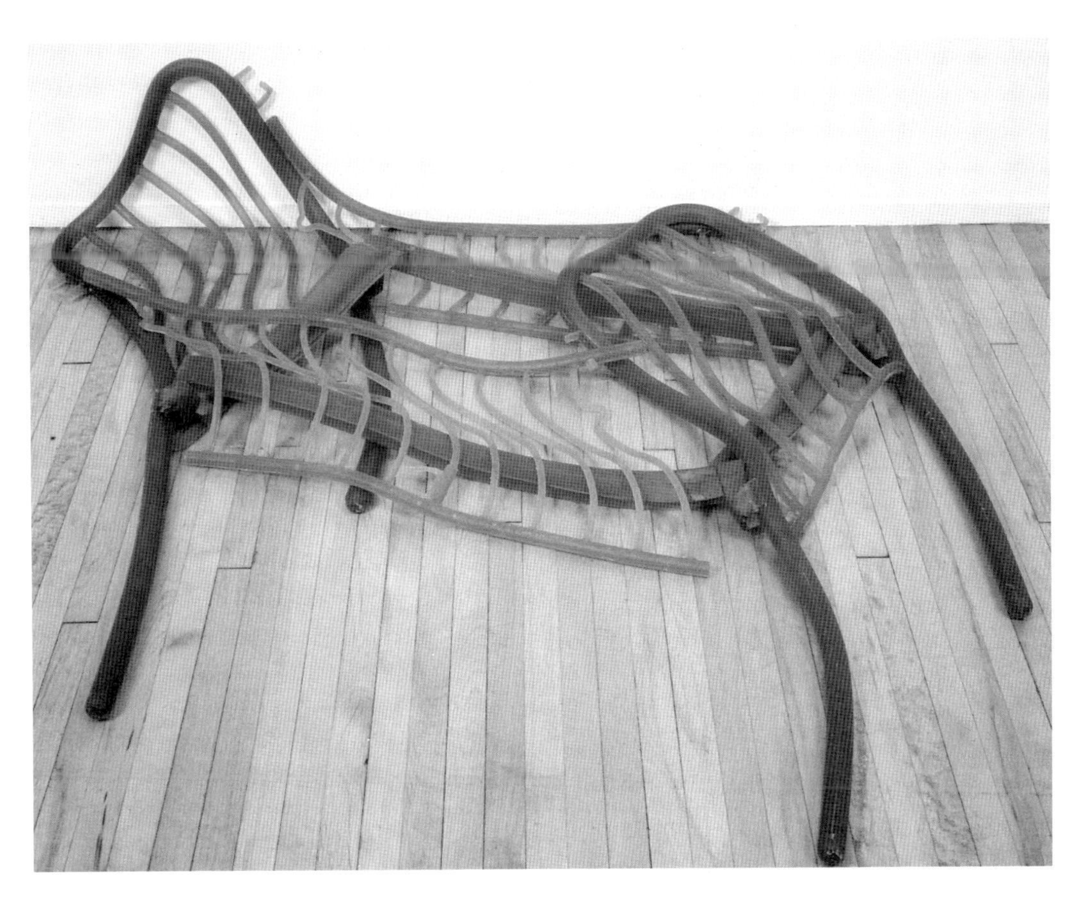

Mona Hatoum, *Marrow*, 1996. Rubber, dimensions variable. Courtesy of the artist.
Photo: Bill Orcutt.

earliest childhood memory, in the body's floppy helplessness and infantile dependency. *Marrow* returns us to the ground – as does *Entrails Carpet* – but from where *being* paradoxically rises. It is an irrefutable law of nature that only the living defy gravity while the dead return to (fall upon) their source. In the implications of the title '*Marrow*' it is as if Hatoum metaphorically links the power of memory to scientific understanding of the stem cells' activity in the marrow from which increasingly specialised generations of cells emerge from a furious genetic switching on and off of 'memorised' signals. In *Marrow*, something (childhood?) seems to have been unsuccessfully shed and like a cast-off skin lies helpless upon the floor. Perhaps, like the Dodo, it is not genetically dead but only waiting upon another generation's ability to restore it to life. The head of one particular Dodo, at present reposing in the University of Oxford's Museum of Natural History, might harbour enough genetic information to produce a clone through the introduction of genetically engineered DNA to existing pigeon stock. In this happy state, its physical body could be up and doing again and *Marrow* resonates with something of the same playfully soft, yet incorruptible resilience of the genes to 'remember'.[29]

Gliding through associations of one shuddery substance to another[30]

In the late Helen Chadwick's scientifically based work, which included her research into *in vitro* fertilisation, she played on more than simply the visual sense and included elements of synaesthesia, the phenomenon in which the stimulation of one sense gives rise to experience in another. Synaesthesia is an interesting word and has an ancient history.[31] The Greeks found the smell of plants and spices highly charged sensations, literally creating atmospheres of sexual freedom or restraint. The modern biological term for the chemical substances secreted by animals, 'pheromones', comes from the Greek *pherein*, 'to carry', and *hormon*, 'to excite'.[32] Ambrosia, the food of the gods, smelt sweet and induced repose, and, according to Hesiod, Zeus seduced Europa 'breathing saffron from his mouth'.[33] Interestingly here, the word 'perceive' is identical to the Homeric Greek for 'I breathe in' while 'aesthetics' is already rooted through the earlier Greek 'I gasp, breathe in'.[34]

Current research into the human infant *in utero* has shown that flavours and tastes consumed by the mother are transferred to the infant's chemosensory environment and that even after birth, these remain plastic and relatively unfocused.[35] With this in mind, we can begin to understand how specific memories of taste and smell can continue to affect adult well-being and how a taste can easily, as with Proust's madeleine, call up sound and colour blended together in the remnants of undifferentiated sense memory. The 'I gasp, breathe in' nature of aesthetic experience therefore becomes more significant when we acknowledge its history in the half-forgotten synaesthesia of earliest life:

> *… and here, undimmed*
> *By any touch, a bunch of blooming plums*
> *Ready to melt between an infant's gums.*
>
> (JOHN KEATS, *Endymion*, 11, 449–51.)

In her work *Of Mutability* (1986), first shown at the ICA, Chadwick contrasted *The Oval Court*, a piece of rococo primavera-like floral loveliness, with *Carcass*, a human-sized column made of household vegetable waste – a compost heap. The power of the olfactory element dominated to such an extent that the gallery was forced to remove *Carcass*. In a later work, *Cacao* (1994), oozing and bubbling chocolate rendered itself soft and presented its sweetness to the air, almost certainly releasing in its mouth-watering efflorescence, those endorphins in the brain which so improve mood and induce calm.[36] Chocolate, as Chadwick reminded us, is a complicated and seductive material in terms of the body. It is one implicated both in Western dietary anxiety and in highly pleasurable palatability. Chadwick 'sculpted' her chocolate into liquid, denying it form and emphasising its olfactory value. Perhaps because it induces such subtle neurochemical alterations, chocolate returns us to our earliest *in utero* pleasures – fluid and osmotic.[37]

Chadwick's last major project before her untimely death in 1996 was an involvement with the *in vitro* fertilisation programme at King's College Hospital Assisted Conception Unit.[38] Chadwick created her *Unnatural Selection* from the fertilised embryos she had permission to use. In IVF treatment, after the surgical harvesting of the prospective mother's eggs, all procedures take place in the laboratory. After two days the fertilised pre-implantation embryos are examined and selected for introduction to the uterus. In her discussion of embryology research, Dr Virginia Bolton raises the intriguing issue of aesthetics in this selection, as the embryologists choose those fertilised eggs with the best morphology and the most rapid division. She points out that there is no clear evidence to support this selection, only that it 'feels' appropriate.[39] Eggs displaying qualities of wholeness and essential vitality are valued above those which are more unevenly formed and/or tardy in their cellular division. Fertilised eggs not chosen for implantation are either discarded or available for research for a period of up to 14 days. Those eggs donated to Helen Chadwick were manipulated by her through suspension in formalin, which terminated their existence even while it preserved them for the photographic work.

In order to manipulate the eggs, Chadwick had to learn the painstaking use of the glass suction pipette. The critic Louisa Buck points out, 'it involved applying a length of tubing to the mouth and, via a series of stoppers and buffers, sucking up the fertilised eggs into a fine filament of glass whilst simultaneously looking down the microscope.'[40] As Buck emphasises: 'Lips, tongue, lungs and saliva paradoxically emerge as key tools of reproductive technology – an intimate alliance which is fraught with visceral, erotic connotations and at first Chadwick found the process hard to stomach.'[41] Poignantly, in some of the 'frozen' animation within the formalin, Chadwick caught the sperm still trying to enter the protective outer membranes of the fertilised and dividing egg. From these discarded eggs Chadwick created the series of photo pieces which include works such as *Opal*, *Monstrance* and *Nebula*. The titles offer many metaphors. *Opal*'s associations conjure the jewel-like surfaces of the dividing cells, while *Monstrance*, with its religious reference to the display of the body of Christ during the Mass, touches upon grief and upon those lost lives which had almost become something or someone. *Nebula* contains all of this and introduces perhaps also the coldest of scientific references. Ambiguous in scale, it can be microscopic or vast, strung out, like a planetary system in space, each heavenly body containing life's individual and vital opportunistic ability to

Helen Chadwick, *Nebula*, 1996. Cibachrome photographs and perspex, panel diameters 46, 60 and 76 cm. Courtesy of the Helen Chadwick Estate and Zelda Cheatle Gallery, London.

grow and adapt. Yet paradoxically, Chadwick hung the cold dark matter of a cataracted un-seeing eye at its centre. With this, Chadwick encapsulated life's chance existence in the vast uncharted darkness of space.[42]

Nothing escapes the net of heaven[43]

The artists discussed above have used the media of photography, video and film, now as commonplace in contemporary art as they are in the technological processes of science. Some artists, however, have engaged with science using the traditional medium of paint, which has allowed, through abstraction, a more philosophical enquiry into its issues. Some of this painting addresses what might be considered aspects of the 'sublime' and it could be argued that in a secular age of technological advance, if there is to be an aspect of the sublime at all it might be experienced at the unpredictable edges of scientific research.

The contemporary American artist Terry Winters has drawn upon research into DNA and the minutiae of living structures to create paintings which address the relationship between biology's innate pattern-making and repetitions and forms of abstraction in the external built environment. Winters has been influenced by scientific theory and, like Orlan, refers to the ambivalent ground of 'inside and outside' in his practice of structurally extending the painting field until it is experienced as a vitality beyond his own body parameters.[44] Referring to such diverse influences as fractal geometry, growth and form, fluid motion and structural systems, Winters reiterates in paint something of the intense energy of these dynamic processes. There emerges in his painting a visually articulated awe of the underlying vigour in all organic states. Forms emerge in the canvases which are never closed down, never completed, and we witness a body in the process of becoming, with a seemingly vast cellular ambition towards increasing complexity. In this respect Winters makes visible what the neurologist Steven Rose defines in his book *Lifelines* as the strange but essential premise upon which all living bodies operate: 'Life demands in all of its forms the ability simultaneously to *be* and to *become.*'[45]

Similarly, in terms of this new scientific perspective on the sublime complexity of the living body, the British artist Therese Oulton has, since the 1980s, created in her work a world of uneasy states and shifting forms. The early paintings almost ooze a viscous and fatty slippage between one vital bodily cohesion and another. The alchemical and religious references and titles of some of her early work have been referred to elsewhere,[46] and there is an element of transcendence in this work which could be compared to the more physical and surgical interventions of Orlan's practice. Oulton's paintings however, as only paintings can, capture a moment at an uneasy borderline between 'flesh' and 'fluid', between 'form' and 'unform', and the result is not so much a chaotic disintegration as a moment in the life of a vital and palpitating, biochemically intelligent flesh. In her discussion of Oulton's *Wishbone* (1988), Angela Moorjani draws attention to the 'hints of genetic process' inherent in the 'fluttering' forms of the painting, which are essentially 'transitional, fleeting, not to be arrested in their flight'.[47]

Oulton's process is continually to build up a structure, sometimes honeycombed and repetitious, while putting that structure under stress in a series of rhythmic, often semi-translucent and layered compressions. This suggests an understanding of the vast profusion and depth of inter-cellular activity functioning in living tissue. In *Precipitate* (1989), for example, there is

Terry Winters, *Gray-scale Image*, 1998. Oil on linen, 243.8 x 304.8 cm. © Terry Winters.

the structure of this flesh's underlying cellular form and, in the same visual field, the liquidity of a convulsive movement across the surface which ripples in a seemingly endless series of peristaltic waves or synaptic discharges. The structural complexity holds together as a substantial physical entity while also suggesting a process of continual movement and change. As Stuart Morgan has written of Oulton's work:

> … the densest forms of material … flout the matter versus spirit debate by containing their own light and even emitting it in their own unpleasant way… Her recent desire to show the outsides and insides of things at the same time, with skin and ribs in full view … constitutes another impossibility which can be demonstrated in painting but nowhere else.[48]

There was a moment when your father died … There is a moment when a single neuron fires in the darkness …[49]

Artists have always had an ongoing engagement with illness and death, even more so recently, since the awareness of AIDS. In his final work, *Blue* (1993), the film-maker Derek Jarman juxtaposed verbal reflections on his illness against the pure but vacant blue of the video monitor screen. The intense electronic blue provides a quality of sublimity which neither completely defies the 'feeling blue' experience which illness imposes, nor limits its association with the sky-blue of infinity. Throughout his personal and candid monologue, interspersed with the voice of the actress Tilda Swinton, which offers in places a dream-like reverie, Jarman counterpoises his physical decline and the misery of his symptoms with the complex pharmacology of the medical treatment; the awful encroaching familiarity of the one with the alien chemistry of the other. The monologue, humane, ecstatic and physically explicit by turns, reflects the mundane and routine awfulness of the traumatic effects of both symptoms and treatment and of the power of memory to ignite uncontainable longing and anguish together. Michael O'Pray, writing on Jarman's films, notes: '*Blue* is a peculiarly difficult film to discuss.'[50] Perhaps the difficulty is that *Blue* does not deal with creative 'as ifs' and 'maybes' but with the authentic prescience of mortality. Few film-makers have dealt cinematically with the certainty of their own dying but Jarman is an exception. *Blue* captures the cruel ordinariness of encroaching decay, while Jarman, its unwilling subject, hurls his words against an increasingly soft and ill-defined target with a final attempt at incisive honesty, and at intervals, a rhapsodic call to an enlargement of human freedom:

> the demented woman is discussing needles, there is always a discussion of needles here … she is having a line put into her neck … the room is full of men and women squinting into the darkness … the side effects of DHPG are … How did my friends cross the cobalt river? … our lives run

> like sparks through the stubble ... Kiss me, kiss me, kiss me again and again ... greedy lips...[51]

Illness has a kind of otherness for which, as Virginia Woolf pointed out, language does not have much vocabulary. Jarman's eloquence lies in his juxtaposition of sound-pictures and the power of the mind to imagine, against the void of *Blue*.

This awareness of the inadequacy of language has been extended in Bill Viola's work to the whole experience of the body as being in the world and particularly to the body's experience *in extremis*. Viola frequently explores through video the disparity between the vast collection of scientific knowledge and individual psychological uncertainty. As Viola has commented, science with its capacity to describe and analyse seems to neutralise events such as birth and death by offering a matter-of-fact approach to them. These accounts, however, leave us unsatisfied when we contemplate the reality. Viola has addressed this discrepancy between science and experience in a number of works, including the 1983 work *Science of the Heart*. In this installation Viola placed an empty but made-up bed in a darkened space with a video screen of a beating heart which covered the wall. The bed itself was spread with a blood-red cover over white sheets. The sound of the heart filled the space and at points in the cycle pounded very fast and then slowed down. Finally, the beating came to a complete stop, until, after a period of silence, the cycle began again. The powerful visual image of the heart exposed in the chest cavity related in a familiar way with the bed as one more container of the human, and this banal site of profound and intimate emotional experience was juxtaposed with objective and clinical medical footage.

Fluids are an important element in Viola's work. The *Nantes Triptych* (1992) moves via three video screens from images of birth (on the left), to that of a suspended and seemingly vital body floating in water (centre), to the image of a woman dying in an impersonal hospital cot (right). Each screen is larger than the human body and the camera, in the last moments of the woman's life, comes up very close to her face, juxtaposing it with the huge image of the baby's face. The woman's dry and heavy breathing is countered both by the cry of the infant and by the continuing sounds of water emanating from the central image of the submerged human body. The spectator is caught within an encompassing cycle where, as Otto Neumaier suggests: 'The singular event may appear tiny compared to the total stream of energy but is, nevertheless, *frightening* to us.'[52]

'There is no answer to birth or death', Viola insists, 'They are meant to be experienced ...' and while he suggests that 'science is ascendant and art is recedent', he believes in the essential truthfulness of the aesthetic experience and its usefulness, particularly in its ability to act 'as a language of the body and avenue of self-knowledge'.[53]

The wonders of minuteness

In concluding these considerations of the body in art and science, I would like to explore the physiological relationship between body and mind a little further. There is still so much in that relationship which is unknown and often artists are amongst the few who have the opportunity and temerity to explore it – or the patient crossing the borders of all their previous experience in a delirium of pain or drug-related physiological change. My own research has taken place as an artist at the Department of Haematological Medicine at King's College Hospital, exploring with patients and staff the different languages used to communicate the experience of illness.[54] Here I have encountered people who have addressed extraordinary boundary shifts of body/self in the treatment of their disease and have felt all of the abject nature and the rich strangeness of an unfamiliar dialogue with the body. For the medical practitioner, the body presents its symptoms and the first casualty of illness is personality. Yet the margin where body and personality combine is a subtle one and for many patients undergoing the invasive treatment which often accompanies serious illness, the body begins to take a more central position within the body/mind dialogue. Bodies swell or become emaciated, hair falls out, the eyeballs grow yellow, even the soles of one's feet become so painful that they cannot be placed on the ground. The body linings die, the gums bleed, the joints swell, until there is complete immobility or there is such fatigue that the body experiences the gravity of Jupiter. One patient described his sudden physical helplessness as being 'like a beached whale' while another explained that during treatment her flesh became so hugely sensitive and prone to bruising that she gave the ward furniture a wide berth, steering round chairs and tables with a surreal level of conscious attention. These experiences remain by-and-large uncommunicated – or reduced to the shorthand of medical notes. The patient's remark of his nurse that 'she wrote down "sick twice" but I was nauseous all day' suggests the inadequacy of language when attempting to capture or record all the subtle nuances of suffering. Generally it is not sufficiently understood that when the body changes, the formerly stable properties of the world change with it. Attractive smells and tastes grow repugnant while particular sights and sounds can trigger the most specific of reactions. One doctor remarked that months after chemotherapy, her patient could not approach the hospital entrance without feeling the old nauseous reaction to her treatment.

While it is common knowledge that fever can induce altered states of perception, less well-known are the effects of drug treatment. In *Blue*, Derek Jarman drew attention to the 'abnormal thoughts or dreams' resulting from one particular drug and the chemotherapy used in treatment for leukaemia can produce similar effects. Chemotherapy 'blasts' are particularly destructive interventions, hovering on the borders between wholesale cell death and cell regeneration, and the body can be pushed to the margin of recuperative ability.

In the resulting dream images of one patient, altered states of scale and physical organisation were experienced. Vast scenarios could be telescoped to minutiae and there was a heightened sense of pulsation, rotation and loss of body boundary:

> I was floating over ... a barren landscape lit by a red sun and sky. I could see herds of animals moving about below and then I'd swoop down and the animals would become people running aimlessly about. Then I swooped even lower and the people became ants furiously working away ... I was plunging into a tunnel made of constantly altering and rotating blobs of colour ... I felt uneasy and alone with no boundaries.[55]

Examples of these altered perceptions of mind/body state in illness can be found in both art and literature. There are also individuals who can recall them. With a sense of the telescopic similar to that described above, a member of an in-depth seminar in analytical psychology I attended some years ago recalled the same reduction of scale during a very serious illness. Describing her experience she recalled 'being smaller than a corpuscle and flowing through my own bloodstream'.[56]

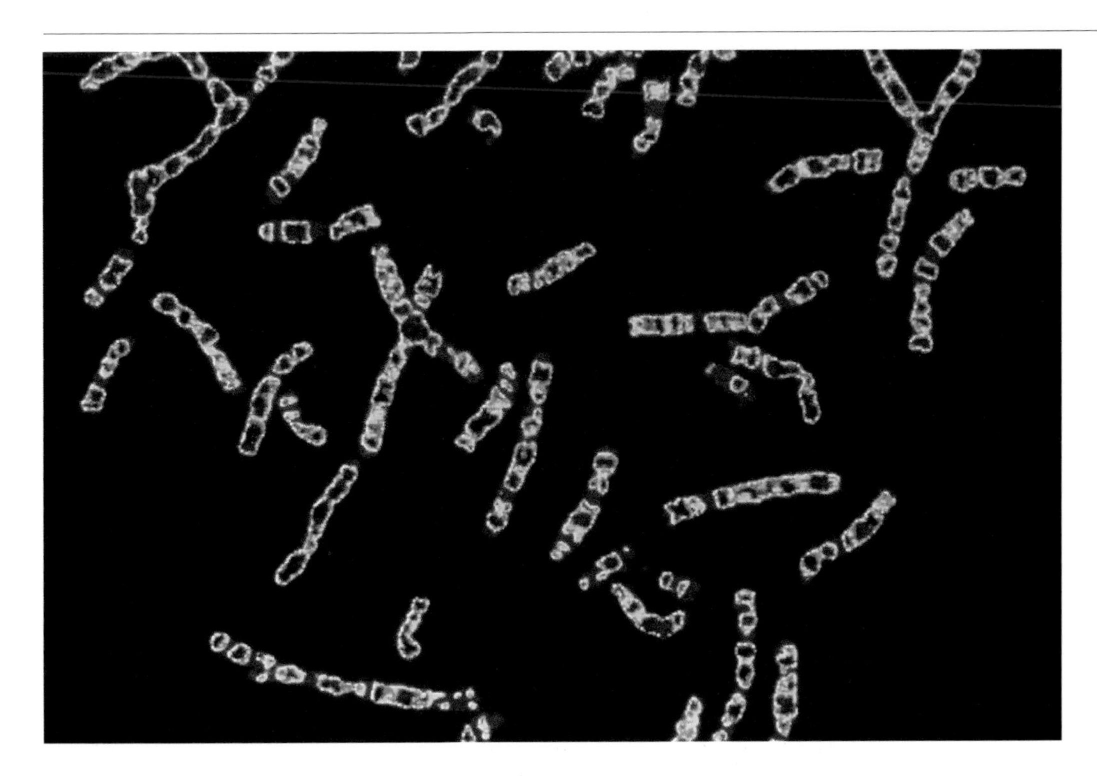

'... of which I was one of the beads' (Dickens, 1853). Human chromosomes. Courtesy of the Wellcome Trust Medical Photographic Library, London.

These experiences have an aspect of physiological/psychological 'mirroring', as though the brain is capable of continuing to function with some sense of 'I' while imaging the body's altered state and its confusion at an abysmal level of physiology. As the brain sinks deeper into its own matter, it seems to be able to maintain a commentary with a symbolising process far down into the teeming activity of the cells: whatever is the matter with 'I' becomes an 'I' which can experience its own matter. While this has partly been defined as the 'Alice in Wonderland' syndrome and is an experience some individuals have – in dreams or the imagination or in drug-induced states – of growing very large or very small, the most extreme 'reductions' seem to take place during illness. A significant example of this in nineteenth-century literature is the account Charles Dickens gives in his novel *Bleak House* to Esther as she undergoes the ravages of smallpox: 'Dare I hint at that worst time when, strung together somewhere in great black space, there was a flaming necklace, or ring, or starry circle of some kind, of which I was one of the beads!'[57] Such a specific description suggests personal experience – either that of Dickens himself, or perhaps related to one of his children. Esther's account, while it seems to have some prescient understanding of the biochemical, is however preceded by a similar one in the eighteenth-century philosopher Edmund Burke's description of sublime reduction, in which, with the same psychological and scientific prescience, he grasps something of the impermanence of scale and boundary – and of the ability of the human to perceive it. 'When we attend to the infinite divisibility of matter … the imagination is lost as well as the sense, we become amazed and confounded at the wonders of minuteness, nor can we distinguish in its effect this extreme of littleness from the vast itself.'[58]

Despite its hard-edged super-realism, much current art concerned with the body has associations with notions of transcendence and this may also be an (un)intentionality within scientific research. Once upon a time, the pterodactyl was real. While scientists rationally ask: how? artists in their more vulnerable and courageous moments may continue to ask the ancient Greeks' preferred question: why?

Here are entities with no fixed identity or allegiance to the ideal, nomadically breeding in an unlimited becoming[59]

What will science next provide for artists? Current research into cloning offers extraordinary opportunities to re-model the body. The next material could be flesh itself – cloned from the artist and perhaps transgenically modified. While science aims to use cloning in order to activate stem-cell production, to produce blood cells, heart tissue or complete new organs, human tissue may soon be grown in any variety – and for any purpose. If Marc Quinn is already using his own blood, why not? The artist's work will literally belong to the artist – the DNA print will be irrefutable – although it could contain a little bit of jellyfish

Marc Quinn, *Self*, 1991. Blood, stainless steel, perspex, refrigeration equipment, 203 x 63 x 63 cm. Courtesy of the artist and the Saatchi Gallery, London.

gene in order to make it glow in the dark.[60] Soon we will all sleep more soundly perhaps, with our own 'jellybaby'. We can therefore feast our eyes on the Chapman Brothers' *Zygotic acceleration, biogenetic desublimated libidinal model (enlarged x 1000)* (1995) and consider that this may not land in the laboratory's bin for incineration despite its aberrations but that, smoothly shopping-mall and techno-glam, this model will emerge ready to party.[61] *Zygotic accelaration* is 'a life-sized fused circle of sixteen bland, genderless child mannequins, wearing nothing but immaculate trainers, their faces obscenely disfigured by misplaced adult genitals'. As critic Louisa Buck observes, 'it presents a chillingly contemporary consumerist spin on monsters spawned by Surrealism's sexual fantasies'.[62] From the specifically post-Freudian context of late twentieth-century thought in which the Chapmans' work sits, the expensive genetic message that *Zygotic acceleration* offers is that all aberrations can become acceptable, as long as they are neat and, like haute couture, without evidence of stitching. While science will offer the artist increasing opportunities to parody and pastiche both existing and extinct life-models, the artist and the scientist may soon come to inhabit the same aesthetic space – one determined by the genes.

Notes and references

1 James Kirkup, from 'A Correct Compassion – To Mr Philip Allison, after watching him perform a mitral stenosis valvulotomy in the General Infirmary at Leeds', in *A Correct Compassion and other Poems* (Oxford, Oxford University Press, 1953).

2 Michelle Hirschhorn, 'Orlan: artist in the post-human age of mechanical reincarnation', in *generations and geographies in the visual arts*, ed. Griselda Pollock (London, Routledge, 1996), p. 120.

3 *Jouissance*: in post-modernist terms is linked particularly to feminist theories of the female body and pleasure. While Julia Kristeva links it to the maternal, Luce Irigaray particularly discusses an 'hysterical' jouissance outside paternal constructions of pleasure and this state relates to qualities of ecstasy experienced by the female mystics (from the Greek *ekstasis*, to put a person out of their senses).

4 See discussion in Hirschhorn (1996), p. 116, see note 2.

5 Orlan, '"I Do Not Want To Look Like": Orlan on becoming – Orlan', in *Women's Art Magazine*, 64, May/June 1995, pp. 7 and 7–8.

6 *Ibid.*, p. 8.

7 Steven Rose, *The Making of Memory* (London, Bantam Press, 1992), p. 8.

8 See Julia Kristeva, *Powers of Horror: An essay on abjection*, trans. L.S. Roudiez (New York, Columbia University Press, 1980); also Antonin Artaud, 'The Theatre of Cruelty (First Manifesto)', in *The Theatre and its Double*, trans. V. Corti (London, Calder Publications, 1993).

9 Orlan (1995), p. 8, see note 5.

10 Parveen Adams, 'Operation Orlan', in *The Emptiness of the Image* (London, Routledge, 1996), p. 154.

11 *Ibid.*, p. 159.

12 Orlan (1995), p. 8, see note 5.

13 Louise Wilson, 'The Electronic Caress: Notes From an Unconscious Subject', in *Touch and Contemporary Art*, ed. David Tomas, Toronto, Public Access, Public 13, 1996, p. 88.

14 *Ibid.*, p. 95.

15 *Ibid.*, p. 88.

16 *Ibid.*, p. 92.

17 *Ibid.*, p. 92.

18 *Ibid.*, p. 92.

19 *Ibid.*, p. 92, drawing upon the research of Johnathan Crary, 'Critical Reflections', in *Artforum International*, February 1994.

20 *Ibid.*, p. 89.

21 See interview with Adrian Dannat, 'Life's like this, then it stops', in *Flash Art*, March 1993, p. 61.

22 M. Archer, G. Brett and C. de Zegher, *Mona Hatoum* (London, Phaidon , 1997), p. 12.

23 *Ibid.*, p. 138.

24 *Ibid.*, p. 138.

25 *Ibid.*, p. 137.

26 *Corps étranger* is a work which raises issues in the work of Julia Kristeva relating specifically to abjection (Kristeva (1980), see note 8) and 'the semiotic'. While digestion raises abjection – the loss of self in the rendering of the body contours in the alimentary canal of [an] other, it also raises the language of the 'semiotic' – the physiological 'dialogue/monologue' between mother and infant which is based on pulsation, rhythm and fluid exchange deep in the flesh and one which is particularly an *in utero* experience. While Kristeva developed 'the semiotic' to define early infant experience it is one which has helped open up the imagination to the experience of complex living systems in general and at levels far below the threshold of consciousness; inchoate, fluid and as driven as it is transitory. For 'the semiotic' see J. Kristeva, *The Revolution in Poetic Language*, trans. M. Waller (New York, Columbia University Press, 1974, 1984).

27 B. Elliott and A. Purdy, *Peter Greenaway: Architecture and allegory* (London, Academy Editions, 1997), p. 53.

28 Archer *et al.* (1997), p. 18, see note 22.

29 For example, see Steve Farrar, 'DNA science could rebuild dead dodo', in *The Sunday Times*, 21 March 1999. Farrar discusses the work of Dr Alan Cooper at the University of Oxford. There has also recently been discussion in Australia of re-creating the 'Tasmanian Tiger' from an infant specimen. This 5 ft marsupial was driven to extinction in Tasmania at the end of the nineteenth century.

30 Marina Warner, 'In Extremis: Helen Chadwick and The Wound of Difference', in *Stilled Lives: Helen Chadwick* (London, Portfolio Gallery and Kunsthallen Brandts Klaedefabrik, 1996).

31 *syn* (together) and *aisthesis* (perception), see discussion in Cretien van Campen, 'Artistic and Psychological Experiments with Synesthesia', in *Leonardo*, San Francisco, 32 (1), 1999, pp. 9–14.

32 Richard Sennett, *Flesh and Stone* (London, Faber

and Faber, 1994), pp. 74–5.

33 R.B. Onians, *The Origins of European Thought* (Cambridge, Cambridge University Press, 1951, 1991), p. 74.

34 *Ibid.*, p. 75.

35 See Gary Beauchamp and Julie Menella, 'Sensitive periods in the development of human flavour perception and preference', in *Taste and Satiety*, Annales Nestlé, 56, 1998, pp. 19–31.

36 See discussion of *Cacao* in Lorraine Gamman, 'Chocolate, Chocolate, Chocolate', in *Women's Art Magazine*, 65, July/August 1995, pp. 8–9. Gamman confides: 'I just plan to keep on consuming chocolate. I do it and I don't know why!' Perhaps the body's vestigial traces of synaesthesia offer a solution.

37 See Adam Drewnowski, 'Palatability and satiety: models and measures', in *Taste and Satiety*, Annales Nestlé, 56, 1998, pp. 32–42. Helen Chadwick, like many artists of her generation, was influenced by the work of Julia Kristeva, but there was more *jouissance* in her art than abjection and, in what can be regarded as her synaesthetic work, a close relationship can be made to the positive aspects of Kristeva's 'semiotic'.

38 This artist in residence project was facilitated by the Arts Catalyst, 1995–6.

39 Virginia Bolton, 'Embryology Research and IVF Treatment', in *Body Visual* (London, The Arts Catalyst, 1996), p. 12.

40 Louisa Buck, 'Unnatural Selection', in *ibid.*, p. 9.

41 *Ibid.*, p. 9.

42 See Louisa Buck's discussion, *ibid.*, p. 10, for her perceptive remarks on *Nebula*. 'The "dead eye" can also be read as a paradigm of the relentless eye of science, a symbol of our desire to probe and examine beyond horizons where sight can normally operate.'

43 Chinese proverb.

44 Terry Winters in conversation with Sacha Craddock and the audience at the Whitechapel Art Gallery, 17 March 1999.

45 Steven Rose, *Lifelines: Biology, freedom, determinism* (London, Allen Lane, The Penguin Press, 1997), p. 142.

46 See discussion in Stuart Morgan, 'Sub Rosa', in *Therese Oulton* (Vienna and London, Galerie Krinzinger and Gimpel Fils, 1986), n.p.

47 Angela Moorjani, *The Aesthetics of Loss and Lessness* (London, Macmillan, 1992), p. 150.

48 Morgan (1986), see note 46.

49 Bill Viola, 'I Do Not Know What It Is I Am Like', in *Bill Viola*, ed. A. Puhringer (Salzburg, Ritter Klagenfurt, 1994), pp.16–17.

50 Michael O'Pray, *Derek Jarman: Dreams of England* (London, British Film Institute, 1996), p. 206.

51 Derek Jarman, in *Blue*, 1993, directed by Derek Jarman.

52 Otto Neumaier, 'Appearance', in Puhringer ed. (1994), p. 27, see note 49.

53 Bill Viola, 'Putting the Whole Back Together', in *ibid.*, pp. 145 and 148.

54 Andrea Duncan, *Humanscape* project, 1998–, in the Department of Haematological Medicine, King's College Hospital, Denmark Hill, London. Funded by the University of East London and the Calouste Gulbenkian Foundation.

55 Gordon Hunt, 'Images of the Mind in Chemotherapy', participation in *Humanscape* project, 1998–, see note 54.

56 See A. Duncan, 'Of Castles and Melts: Dickens and the Dark Sublime', in *On the Sublime: In psychoanalysis, archetypal psychology and psychotherapy*, ed. P. Clarkson (London, Whurr Publishers, 1997), note 65.

57 Charles Dickens, *Bleak House*, 1853. Quoted in Duncan (1997), p. 209, see note 56.

58 E. Burke, *A Philosophical Enquiry into the Origin of Our Ideas of the Sublime and Beautiful* (1757), ed. J.T. Boulten (Oxford, Blackwell, 1958, 1987), p. 73.

59 K. Biesenbach, and E. Dexter, in the foreword to *Chapmanworld*, Dinos and Jake Chapman, David Falconer, Douglas Fogle, Nick Land (London, ICA Publications, 1996).

60 This refers to recent research at the University of Hawaii, Honolulu, in which mice were given an additional fluorescent gene from jellyfish. When lit by ultraviolet light, the mice glow green.

61 Discussion at the Natural History Museum, London, on 8 April 1999, between Dr Per Ahlberg and a group of some 200 specialists from the international field of evolutionary science. This addressed the issue of 'evo-devo' and the importance of genetic mutations in the genetic codes of existing life-forms for development of new forms and functions in living organisms. Unfortunately, mutations also give rise to cancers and provide viruses with their often lethal abilities to evade mapping. For further information on the work of the Chapman Brothers, see Chapman *et al.* (1996), see note 59.

62 Louisa Buck, *Moving Targets: A user's guide to British art now* (London, Tate Gallery Publishing, 1997), pp. 72–3; almost as shocking, however, is the photograph of the two-headed dog created by experimental surgeon Vladimir Demikov, which appeared in *Life* magazine in the early 1960s (recently reproduced in the *Times Higher Education Supplement*) and the 30 head transplants on rhesus monkeys conducted by American surgeon Robert White who is proposing to undertake human head transplants (Tim Cornwell, 'Your head in his hands', in *Times Higher Education Supplement*, 29 October 1999).

9 Fission or Fusion: Art, Science and Society

Siân Ede

And I remember the press full of doctors,
of inventions; a herringbone fragment
of DNA to fool a virus, a wisp
of vitamin to lock on to inner decay
and knock it dead for good.

JO SHAPCOTT, from 'Delectable Creatures'.[1]

At the beginning of the last year of the millennium the *Times Higher Educational Supplement* asked nine leading scientists to predict the big scientific breakthroughs of the twenty-first century. Richard Peto, professor of medical statistics and epidemiology at the University of Oxford, foresaw 'the discovery of a Theory of Everything' and '80–90 per cent of the world's population living to 70'. Richard Dawkins, Simonyi professor of the public understanding of science, also at Oxford, predicted, 'the full understanding of human consciousness'. Sir Aaron Klug, president of the Royal Society, was in agreement, his prediction being, 'comprehending the workings of the human brain'. Charles Baker, head of fusion research at the University of California, San Diego, anticipated 'clean fuel', and Roger Angel, professor of astronomy at the University of Arizona, said, 'searching for extra-terrestrial life'. Sir Robert May, Chief Scientific Adviser to the Government, foresaw 'a fuller understanding of genetics', with Lee Silver, professor of genetics at Princeton University, specifying, 'the genetic engineering of children'.

Sir Harry Kroto, the 1996 Nobel-winning chemist, refused to entertain the idea, saying, 'if you can predict a discovery, it can't be that important. Major advances should be totally unpredictable.' Only Susan Greenfield, head of the Royal Institution and professor of pharmacology at the University of Oxford, foresaw 'the engagement of the public in science and the expression of scientific ideas in a way they can understand and contribute to'.[2]

Two of the visions suggest material benefits. A greatly increased life-expectancy world-wide must surely be an attractive idea and clean fuel might be entirely desirable, although it should be remembered that the writer is head of fusion research. But is Lee Silver deliberately taking on the mantle of Dr

Frankenstein when he talks of the 'genetic engineering of children'? He goes on to say:

> The breakthrough will have an unprecedented impact on the practice of medicine, but its impact on the way people have children will be even greater as prospective parents (who have enough money) are given the opportunity to choose the precise genetic makeup of their children. Ultimately the ability to control the genetic makeup of children will change the nature of the human species.

He does not tell us whether he thinks this is a good or a bad thing. It will simply happen. No wonder the public is ambivalent about scientific 'progress'.

All the above prognostications take scientific advance for granted and no one questions whether it is inevitable or whether it may have reached the peak of its success and be currently, as some dissenters believe, 'moving at maximum speed just before it hits the wall', or at least slowing down after a couple of centuries of breakthrough.[3] Like Prometheus, like Pandora, like Icarus or Eve these scientists simply dare and as seekers after knowledge they appear to operate on a mythical scale outside any rooted and recognisable location of time and place.

Lewis Wolpert speaks for many in the science world in his assertion that 'science is value-free', and this idea is worth closer analysis.[4] The statement requires an awareness that the word 'Science' has at least four discrete meanings. Science can mean knowledge itself, which suggests a conflation of the subject known and the knowledge we possess of it. Science is also a method of discovery, an intellectual endeavour which draws on agreed rational processes. The word is used to represent the scientific community at large, and appears to suggest that this community acts as one, in total agreement. And it is used to refer to a technological application of the discoveries of science to the activities of the world we live in. For Wolpert's statement to be other than shocking, it must surely refer to the first two definitions only, the knowledge itself and the rational methodology used to discover it; and, intellectually at least, you can see what he means. It is a thing apart, it is as it is, it is not good or bad, it is 'free'. If you are a phenomenologist or a cultural relativist, and question either the existence of reality itself, or the possibility that it can exist independent of human considerations, you will not be convinced by this belief. You may, however, be able to recognise that it comes from a benevolent desire to provide an agreed system of judicious and impartial knowledge to stand against distortion, ignorance and abuse. This is the Enlightenment dream which places its highest value on finding out about the world by applying unbiased judgement based on a sound appraisal of the available evidence.

If one could accept the benevolent imperative all might be well. But a belief in the impartiality of science has its negative side, for if the scientific enquirer is unable to imagine the potential consequences of his research, then, locked in the laboratory as in popular myth, he may find his pursuits leading him in dangerous directions. It is no surprise that the phrase, 'It would be very interesting to see what would happen if I added a millilitre more to the mixture ...' has featured in so many B movie mad scientist scenarios. Add to this the not uncommon observation that those with vested interests are able to buy into science, for profit, or power, or worse, and it is no wonder that the public is reluctant to disassociate it from value judgement at any stage.

If some scientists refuse to engage in value judgements in their working lives, then the rest of us must. It is heartening to note that Susan Greenfield would like to see the public 'contribute' to science, because some scientists seem to suggest that if only the public understood better, all would be well. They appear to rule out the possibility that if the public understood they might not want to support the enterprise at all, or only selected parts of it. The public is not at ease with judgements which arrive at a conclusion 'beyond all reasonable doubt' when the 'reasonable' element depends on a statistical analysis of likely risk. A Department of Health publication of 1997, *Communicating about Risks to Public Health,* lists eleven indicators which are likely to give the public concern. While some may suggest an over-cautiousness arising from simple unfamiliarity or even from a sentimental view of the 'naturalness' of nature, such as risks 'which arise from unfamiliar or novel sources' or those which arise 'from man-made rather than natural sources', others are justifiably worrying. These include risks which are perceived to 'cause hidden or irreversible damage' and those which 'pose some threat to future generations' and especially of concern are those which appear 'to be poorly understood by science' and those which are 'subject to contradictory statements from responsible sources'. A number of current controversies come to mind – the genetic manipulation of organisms, the commercial patenting of genes, the cloning of animals and even humans, the disposal of nuclear waste – and there are many others. For scientific 'progress' has a less than perfect track record and we may be right to be wary, even while we give some advances a cautious acceptance. Ethical and moral decisions are difficult, intricate and never fixed and require all of us to apply our best powers of reasoning and our emotional instincts to the task, keeping us as wary of politically motivated scaremongering as we should be of over-patronising reassurances. And we shall clearly need all the energy we can muster to worry away at the ethical considerations arising from the kind of genetic engineering envisaged by Lee Silver.

We saw in Chapter 2 that the highest ideal for the communities of both science and art was the notion of freedom. For scientists the scientific method is the most straightforward route to a dispassionate 'truth' in defiance of superstition, delusion and exploitation. For contemporary artists, freedom is more likely to be about dissent and claims to a privileged truth are distrusted. Debates and arguments about the nature of freedom are stimulating and need to continue for as long as any form of freedom exists, but if they are too theoretical they sound like fiddling while Rome burns. The struggle for impartiality is not an easy one, especially where other interests predominate and are able to use persuasive means to cajole us into apathy or false security or, alternatively, panic. We must all be alert. With particular prescience, the playwright Henrik Ibsen dramatised the fight for an independent and impartial view as long ago as 1882, in *An Enemy of the People*, where the hero is a scientist who is made a public pariah because he points out that the spring water upon which the town's economy thrives is contaminated.

Just as the human mind can hold many forms of knowledge at once, so can society accommodate the different approaches of science and art when it comes to maintaining freedoms. There is a place for narrow reductionist enquiry as long as other people are allowed to see the context as a whole and have equal status in determining how to use knowledge to the benefit of all. There is a place for materialism, a quest to understand the basis of the matter of the universe, as long as there are others who can consider the metaphysical implications of this for our systems of thought and belief. Reason, logic and evidence-gathering are recognisable instruments of justice provided they are exercised in a context where there is a spirit of tolerance and integrity. Artists, like the rest of us, must learn the new language necessary for an understanding of science, engaging even in highly technical methodologies. For if we do not begin to understand the culture, we will not be respected for our grasp of the issues involved and will lay ourselves open to being ignorant or sentimental or trivial, able at best to make playful but unhelpful distortions. Scientists must be persuaded to view their work, at every stage, from the perspective of others. They should encourage the 'public understanding of science', certainly, but ideally we should work towards a 'public consultation on science'. If we can work as equal partners we should be able to bring our intelligence and sensibility and our common quest for freedom to bear on politicians and opinion-formers so that we are free from the manipulations of vested interests.

In the remainder of this chapter we will look at four different examples of ways in which artists have engaged with the repercussions of scientific discovery and practice. On some occasions the artists have been subtly critical. On others, however, the artists have collaborated with scientists but have taken a personal route in order better to understand the work in hand, showing how we may all consequently play a role in making informed decisions. These interventions are important indications that artists can make a real contribution to a benevolent development of science in the twenty-first century.

The cat's pyjamas

American artist James Acord creates granite sculptures like medieval reliquaries in which he places radioactive material. If this appears to represent a shocking subversion of the sacred, his intention is not a laboured attempt at irony. 'I am simply a sculptor,' he says, 'I live in the nuclear age, so it is logical, perhaps even inevitable, that a sculptor would work with radioactive materials. I'm not for or against the nuclear age, but we are all in it, and we should be dealing with it.'

Although Jim Acord is besotted with the transformation of materials and the artwork itself is of the greatest importance to him, the story surrounding the work is essential to understanding its political and metaphorical implications – and they are not simplistic ones. Standing in a lecture theatre in his regulation grey suit and army-style haircut, but with his nuclear licence number tattooed on the back of his neck, he is in a class of one – a unique creature, neither wholly artist nor scientist – and even while he is simply telling you the facts, he is a compelling entertainer. Seattle-born Acord's fascination with nuclear materials began when he realised that the granite he habitually used for sculpture is the most radioactive of all stones. This realisation started him on a trajectory which ended with an MA in nuclear engineering and ambitions for a series of artworks containing this most dangerous of materials. He learned that granite deposits were habitually used as possible sites for the long-term storage of nuclear waste and planned a sculpture called *Monstrance for a Grey Horse*, inside which he wanted to place radioactive material. He visited ore museums and discovered in flea markets a fancy type of tableware with a bright orange glaze which contained uranium. He would buy it and, wearing protective clothing, chip it off and pulverise it. It was not long before his activities came to the attention of Washington state officials and it took two years of negotiation to get permission from the US Regulatory Commission for him to handle the material. As his passion developed he visited the Hanford Site, the largest atomic facility in the West, which had produced the plutonium used in the Fat Man type A-bombs which were dropped on Nagasaki in 1945. Constructed in a hurry during the war, the site had become the most contaminated nuclear area in the US, with leaking underground tanks, although it also had a new sodium-cooled reactor, a Fast Flux Test Facility. The lure of this inspired Acord to move permanently to Hanford's dormitory town, with a view to getting access to the reactor because of its capabilities in transmuting one metal into another. Attending classes at Hanford's graduate education centre he found himself fascinated by nuclear physics. 'I realised that the transmutation of elements, although an age-old dream, is a reality since we've entered the nuclear age,' Acord reported. 'The ability to manipulate one elemental substance to another is the cat's pyjamas. This is what artists have dreamed about.'

His studies advanced to classes in nuclear engineering so that he could acquire a licence to handle radioactive material, the only artist ever to have achieved this, and at Hanford his main ambition remains to build a massive henge-like monument, incorporating the twelve breeder blanket assemblies donated to him in 1993 by Siemens, the only company prepared and permitted to do so, after a great deal of sensitive negotiation.

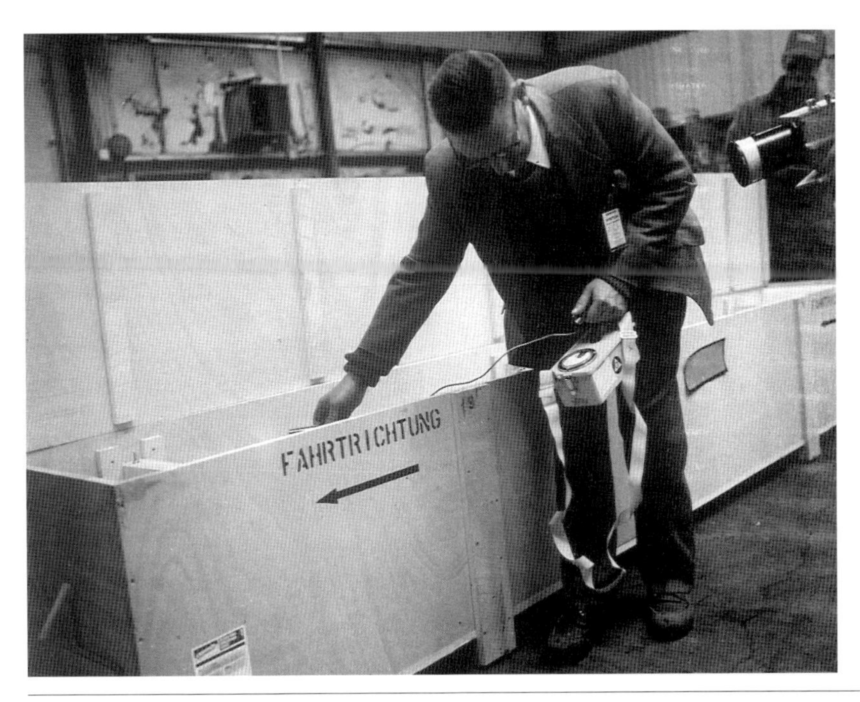

James Acord inspects radioactive uranium breeder blanket fuel assembly rods delivered to the Hanford Site, 1993. Courtesy of James Acord. Photo: Arthur S. Aubry.

In 1998 he was invited by the Arts Catalyst to undertake a residency at Imperial College so that he could work with physicists in adopting the process of fast neutron capture in order to transform milligrams of an isotope of technetium, derived from an extract of nuclear waste, by 'bumping' it up to the next element on the periodic table, ruthenium, which is a stable non-radioactive member of the platinum family. Technetium is difficult to work with because it emits penetrating gamma rays and has a half-life of 22,000 years, but its transformation was eventually achieved. The resulting substance was placed in wall-mounted granite bas-reliefs, their traditional memorial-stone appearance reminding the viewer of the artist's abiding obsession with the transformation of materials and its ancient link with alchemy and ritualised religion. In this contemporary version the profane is rendered sacred.

The quotation chosen by writer James Flint as a preface for his exhibition catalogue essay about Acord reflects on the almost sacred vow of secrecy required by scientists who take part in the nuclear industry:

> 'We had voluntarily accepted a measure of secrecy', wrote Robert Weiner, 'for the sake of the war ... We had hoped that this unfamiliar self-discipline would be a temporary thing, and we had expected that after the war ... we should return to the free spirit of communication ... which is the very life of science. Now we found that, whether we wished it or not ... At no time in the foreseeable future could we again do our research as free men.'

As an artist, James Acord has revealed some of the mystique to the rest of us. The nuclear industry is a source of grave disquiet to the public and it is probably true to say that is deeply reviled by the many in the arts community, when it is not banished cursorily out of mind. Acord's brave and honest spirit of enquiry, which never loses its sense of artistic purpose, meets science with science, and attempts to throw the light of integrity on an area of science which has been unable to view itself with openness, almost since its inception. The artwork as a whole – the sculptures themselves, Acord's story and the performance and pictures of it – provokes in audiences a disorientation of values and they must find their own meanings. Acord prefers to stand on the sidelines but will admit:

> I genuinely hope that my artwork can make a contribution to resolving some of the problems of what to do with high-level radioactive waste. I get a lot of flak over this from both sides. I'm not proposing this art as an answer to the problem, but it is a symbol and a metaphor for the fact that we do have control over this. We have created these elements. We can uncreate them. [5]

The future's mirror

Artist Cornelia Hesse-Honegger makes exquisite aquarelle paintings of jewel-like bugs and flies but, unlike James Acord, she has found herself at the centre of an awkward controversy affecting the nuclear power industry. Hard-edged and delicate at the same time, glowing with gem-like colours, finely detailed with spots, hairs and markings and presented either as single specimens or as apparently ordered taxonomies, the enlarged images of microscopic observation attract the viewer's eye as objects of beauty. At first glance all the pictures reflect nature's perfectibility, until the eye is drawn to those which present strange asymmetries, areas of minute distortion, crumpled wings, deformed heads, blistered eyes and abnormally coloured growths. When viewers first learn that the bugs have been collected down-wind of nuclear power stations or in various European countries in the wake of the Chernobyl disaster, they are likely to experience a visceral unease. It is clear to see why the artist has been accused of sensationalism.

Hesse-Honegger was a professionally trained scientific illustrator who worked in the Department of Zoology at the University of Zürich in her native Switzerland, where her main task was illustrating scientific publications on taxonomy and laboratory-induced mutations of Drosophila, the fruit fly. Drosophila are customarily used for genetic research because they reproduce quickly and patterns of mutation, accidental or deliberately induced, can be easily observed over many generations. When she left work to bring up her children she continued to collect and paint bugs for her own pleasure and

Cornelia Hesse-Honegger, *Harlequin Bug* (Pentatomidae), 1991. Deformed scutellum and asymmetrical patches. Found near Three Mile Island nuclear plant, Pennsylvania. 12 x 20.7 cm. Courtesy of the artist and Locus+, Newcastle upon Tyne. © DACS, 1999. Photo: Peter Schälchli.

found them abundant around her garden. But she began to get worried when she started noticing a marked decline in the numbers of some species at the end of the 1970s, going so far as to report her concerns to the Department of Zoology in Zürich. Her interest was properly aroused when the Chernobyl disaster occurred in 1986 and she assumed that the widespread contamination resulting from the nuclear fallout over many European countries would provide scientists with an outdoor laboratory of enormous size and interest. She was somewhat surprised, therefore, to be met with little more than mild reassurances from the scientists she talked to and so began her own research, visiting Sweden, which had had the heaviest fallout in Western Europe, a year after the disaster, by which time any effects might be visible in subsequent generations of insects. She collected specimens and was alarmed to observe a high proportion of deformations. She went on to the area of Switzerland which had received the highest fallout and started showing her work.

Hesse-Honegger was unprepared for the criticism and anger she received from the scientific community, which had been reassuring the public that in general the fallout doses were low and would cause no particular harm. She widened her research to areas down-wind of some of the Swiss power stations, and particularly to places where there were supposedly safer, low levels of radiation, but where she found that insect mutations were almost as numerous. In 1989 she visited Sellafield in Britain and in 1990 joined a party of Swiss parliamentarians and journalists to Chernobyl itself. She went on to Three Mile Island, the notorious Pennsylvania site of a near-meltdown in 1979, and continued her research around power stations in Switzerland. By this time her work had become well known and she was applauded by many in pressure groups but reviled for her apparent travesty of the scientific method by the orthodox science community.

It would be simplistic to say that she had disturbed a conspiracy of silence, motivated by an industry which, as Robert Weiner's quotation showed us earlier, has always had to operate in the confines of secrecy or by governments which could not bear to contemplate the consequences – and costs – of admitting that there was a genuine environmental problem. It would be difficult to prove whether some in the scientific community were working hand in glove with the industry and the governments, as some of Hesse-Honegger's supporters suspected. But what appears to be at the root of the questioning of her work is more than simply political expediency. Scientists in general were sensitive – and still are – because she had appeared to abuse and distort the scientific method. Unless she could be seen to follow the strict protocols established for impartial investigation, her results would be scientifically meaningless. The scientists' main objection was that she had not used a biotope, a natural place with a specific formation of plants and animals supporting its own distinctive community, which could stand as a control against which she could measure her new results. She did, in fact, make use of such a place – the Swiss Alps, where there are no nuclear power plants – but fundamentally she was in disagreement with the principle, on the grounds that it would be impossible to locate anywhere on earth uncontaminated by pollution from half a century of nuclear experiment. While it would, indeed, be difficult to find any prelapsarian paradise, her uncompromising attitude was to undermine the impact of her findings from the point of view of the science community. She had failed to produce sufficiently detailed taxonomies or scientifically accurate analyses in terms of samples, numbers and percentages with comparisons and controls measured over significant periods of time. It was simply not enough to say to scientists, 'all I can say is that when I went to the Swiss and Italian Alps the bugs I found looked healthy and fine.'

Scientists gained the impression, therefore, that she was working to a preconceived emotionally or even politically motivated agenda and they were quick to point out the inadequacy of her data and her failure to acknowledge properly, with accompanying statistics, that mutations and deformities are normal in nature. And even some environmental scientists sympathetic with her aims were not happy with the uneasy blend of art and science, of emotion and 'evidence' in her work, for it appeared to parody where it was trying to convince and was therefore rendered less credible. Some even felt she had weakened their own cases, carefully built up through respecting the orthodox scientific protocols.

However, if scientifically inadmissible, her efforts were brave and there is no reason to believe that she was motivated by other than the highest integrity. She was one artist, working on her own, and the size of the study required for a proper scientific analysis would have demanded enormous resources. If the scientific community was doubtful they could, after all, have proved her wrong with a big, properly resourced study of their own. For it is certainly true to say that her work successfully raised the public's awareness of a situation which they, like their governments, would prefer to see disappear from sight and mind. The paintings and the accompanying 'story' became widely known and hotly discussed. And Hesse-Honegger did have limited success politically in instigating a few new scientific studies within confined parameters, although the scientists concerned were not able – or, in her opinion, willing – to point to such dramatic results.

In 1992 the Swiss Federal Office of Culture presented an exhibition and book of her work as the Swiss contribution to the Milan XVIII Triennial. This subsequently toured to several places in Switzerland, to Vienna and Sweden as well as to British cities in 1996 and 1997. It is chastening to note, however, that their success as artworks may have actually relegated them to an area of secondary status. Confined to galleries they occupy an arena where they do less harm and instead of representing a millstone of responsibility they are transformed into national emblems of high culture.

Hesse-Honegger set herself an ambitious task, driven by her regard for the natural world although, in the beginning at any rate, and despite her professional experience of the science world, she was perhaps naive about the reception she thought she would get. Nevertheless she has raised public consciousness. Her pictures featured as a cover story on *Time* magazine in the US, the debates she aroused were passionate on both sides and anyone who has seen her work cannot fail to go away with its images imprinted on their mind. She writes, 'If the only way to stop the daily disaster is to frighten people then I am happy if my paintings help', and if this sounds too provocatively emotional for some critics they must acknowledge that she has followed her conscience in very difficult circumstances. It is now up to the viewer to consider whether the artwork can operate as little more than sensationalist propaganda, undermining the very case it sets out to make because of its less than thorough methodology, or whether it has effectively revealed information to cause real concern. Certainly it has failed to convince those who need most convincing. At the very least, however, it has drawn attention to an uncomfortable possibility which many people would rather not contemplate and it may urge them to find out more, to ask pertinent questions and to take democratic action wherever and however necessary. It has also reminded us in general – for we keep needing the reminders – that the environment is precious and in danger and, as many respectable scientific surveys reveal, it is an issue which is urgently in need of being addressed.[6]

A frolic of information

The artist's residency at the UK Human Genome Mapping Project (HGMP), near Cambridge, was set up by Trystan Hawkins of the Wysing Arts Centre, Cambridgeshire. Wysing, an eleven-acre farm converted into a gallery with studios, is keen to develop relationships with its industrial neighbours. A two-year project, called *Wider View*, was advertised nationally and two artists were selected for the first year. Natalie Taylor was chosen to work with horticulturalists at Writtle College, and Neal White was to work at HGMP.

The HGMP Resource Centre is involved in the international race to map the entire human genome and it provides genetic material to scientists worldwide. It also runs a research programme on genomics and disease. As in most science establishments the pace is brisk and preoccupied. The laboratories are pristine and well-organised and contain elaborate machines. Staff appear to be engaged in measuring things but there is little for the outsider to see. The real work happens on computer screens and, as the artist was to learn, in the data, minds and discussions of the investigating scientists.

Although he has no formal scientific background, Neal White came well-equipped to cope with the technology. He had studied for an MA in computer programming and fine art and has a practical attitude to work, having co-founded a group called Soda, a digital technology consultancy which generates revenue in order to subsidise the art he and his colleagues really want to make. Nevertheless, as an individual in a highly organised and frenetically busy establishment, he would have to make a niche for himself, forge the right relationships and learn to understand complex scientific processes in order to come up with at least one work of art, an individual and sensuous response which would be understood and liked by his hosts. On his first visit he made inroads into the task. He was given formal security clearance and a space to work. By the end of the day he was clutching guides to genomics, had signed up for a course on bio-infomatics and was deliberating on the potential for the making of art in such an environment. White does not describe himself as a 'digital artist' but as an artist who happens sometimes to use digital technology as a medium. The scientists at HGMP generate huge amounts of data, looking for consistent patterns in chromosome sequences and translating them into qualitative information. There was obvious potential for displaying pattern for its own sake, but the overwhelming function of the HGMP relates, of course, to human life and it is the conceptual implications which make the research so intriguing and frightening to non-scientific observers. 'If science can deconstruct and reduce the identity of the individual to a set of data, can it also equip us with the facilities, references and validations by which we can make sound moral and social judgements as to how that data is used?' asks White.

In the event, he chose three ideas under the general title *Inheritance (Hotel)*. In the final display at the converted barn gallery at Wysing the viewer first walked down a narrow corridor beguiled by the sounds of voices and by still photographs on the walls. The corridor opened into the main space and the viewer faced a large screen. Panels of text, which the viewer could read or ignore, occupied some of the other wall-space.

Inheritances

The attraction of HGMP's work for the artist arose from the fact that it encompasses and is driven both by advances in technology (the computational ability to cope with vast amounts of information) and by developments in scientific technique, such as Polymer Chain Reaction. In *Twentieth Century Screen*, a piece White created for The Lux Gallery, London, in 1998, he had utilised advances in electrochemical technology and this inspired his interest, first in chemistry and then in biochemistry. Polymer Chain Reaction (PCR) is a technique which represents an assimilation of technology and scientific method and it enables scientists and laboratory technicians to generate unlimited copies of any fragment of DNA. It is an advance which has truly revolutionised the remit and scope of the work carried out by geneticists in the biotechnological era. White was intrigued by the fact that this vast information-gathering is rendered objective and anonymous, yet paradoxically is concerned with issues which can have meaning for the non-scientist only in a deeply personal sense.

As it happened, the artist had a moving personal story of his own to relate to his work. At the age of fourteen he was told that the grandmother who had lived in his home all his life was not a blood relation: his mother had been adopted. Some years later he persuaded his mother to find out about her 'real' or 'biological' mother. She went to the adoption agency to read her adoption file but as she opened the file and read about the circumstances which had led to her adoption she was presented with a photograph of 'your adopting mother seeing you for the first time'. 'This simple picture,' writes White, 'the image of a woman leaning over the child she has yearned for, would have been very powerful in itself, were it not for the fact that *the statement was not true*.' The picture bore no resemblance to her adoptive mother and was, in fact, a picture – the only picture – of her real mother, which had been wrongly filed many years before.

For the first of three artworks, *Inheritance 1*, White sent a copy of this story to each member of HGMP's staff and asked each for three personal photographs – a picture of themselves with two other family members, grandparents, parents, siblings or children. Not everyone was prepared to submit tokens bearing such deep personal resonance but enough photographs were collected to be presented, digitally enhanced, in an automated slide sequence at the exhibition, through a dark and almost secret hole in a side wall.

Inheritance 2 concerned an even more dramatic application of genetic knowledge. The HGMP Resource Centre hosts the oddly named 'Linkage Hotel', a facility where doctors, clinical and medical researchers can establish links between family members who may be affected by hereditary illness, and

Neal White, *Inheritance 1*, 1999. The artist as a child (on the left) with his brother and 'adopted' grandmother. Courtesy of Neal White.

where genetic counselling is also provided. This project was created in collaboration with clinical genetics counsellor Evan Reid who conducted with White a mock counselling session. 'The genotypes, which form an informational identity of these people, represent more than just a string of numbers or code,' Evan Reid told the artist. 'It's not just a matter of looking at the sequences. I have met these people, I have heard their concerns, I know their fears, I know who they are, I have heard their voices.'

The resulting artwork is in the form of an audiotape of the fictional counselling session, written with the writer Lawrence Norfolk. Inconclusive and hesitant, the dialogue is apparently naturalistic but made artificial by the deliberately heightened articulations of the actors speaking the lines. We are ill-equipped to discuss matters of terrible import and ordinary words can thus become weighted with poignancy.

Inheritance 3 is the most high-tech piece and the one which most closely reflects the nature of the scientific research at the HGMP Resource Centre. A low-resolution photographic image of the artist is divided into 380 discrete pixels. The number of pixels corresponds directly to the number of markers contained within the artist's genotype, which was established after a blood sample was refined to pure DNA and marked by HGMP researchers. As a controlling computer causes the pixels to become illuminated one by one, the genotype is gradually revealed, made human as a photo-image of the artist's face. At the completion of the image a somewhat bizarre additional event is triggered in the firing of a sound cannon (bird-scarer) located outside the gallery.

The finished works, experienced as one moved through the gallery, were striking in themselves but even more so, when given the background information. One of the unexpectedly successful consequences of the project, however, has been the strength of the relationships forged between the artist and the staff at HGMP. The staff had hoped that the eventual artworks would

Neal White, *Inheritance 3*, 1999. Pixelated image, 183 x 183 cm. Courtesy of the artist and Wysing Arts, Bourn.

assist with increasing the public understanding of their work but they had not realised how they themselves might learn to understand art better. Scientist Michael Rhodes writes:

> It was with some trepidation that I approached this project, but although there is still art out there that I don't personally like, I now understand what the artist is trying to do. The amazing revelation for me is that art should not be explained. As a scientist, part of my training is presenting and explaining ideas. I do want my audiences to think about the implications of what I discuss but this should be based on an understanding of the subject. An artist creates a work to provoke thought but as long as the audience is engaged by the piece there is no correct way of interacting with the work. A scientist explains: an artist provokes a response. Science has answers (hopefully): art has no limitations. It was really interesting to see Neal learning about the biology of inheritance and seeing the understanding alter what he was creating.

Beyond the exhibition of artworks, this new relationship has a potential for further development and Neal White has been invited back to HGMP to continue his research there and to develop other collaborative projects. He writes:

> The science of genetics implies a sense of playing the odds, of gambling. The very fact that any of us is even here is, of course, a matter of chance. The scientists at HGMP believe that genetics holds the key to many of the hands dealt to us. However, if the props we have come to rely on feel inadequate in helping us deal with the problems and issues raised by the human genome project, and our moral and logical capacities are challenged, then we may feel that our only route is to accept the wild cards of fate and destiny.[7]

The seen and the unseen

The evocatively named Durham pit village of Quaking Houses has been fighting for its existence since the Billy Pit was closed in the 1950s and the village was left to die. The surrounding land was subsequently selected as the site for Durham's largest opencast development, Chapman's Well, a deep excavation in the pattern of an inverted question mark drawn around the village. When the developers succeeded in applying for a substantial extension to this site and in response to the industrial blight all around them, the residents formed the Quaking Houses Environmental Trust. Their first objective was to reclaim the barren waste land where the Billy Pit had stood. But the evidence of minewater pollution, particularly in one local stream or 'burn', had become a further concern.

Minewater pollution is one of Britain's most serious causes of environmental degradation. The oxidation and dissolution of the mineral pyrite, which is widespread in deep coal mines and spoil heaps, causes bodies of water in contact with it to become charged with heavy metals, iron and aluminium in particular, and these waters, which may also be acidic, ultimately discharge into surface streams, contaminating the environment. The immediate and unsightly consequences are heavy deposits of orange-red iron precipitates on stream beds. When high concentrations of aluminium are also present, the deposits of iron 'ochre' are contrasted with floating masses of white froths, suspensions and precipitates. Even if the resultant ecological destruction is not complete, it is always severe.[8]

This pollution had also come to the attention of civil engineers at the University of Newcastle, and a postgraduate student had independently presented a report on the state of the Stanley Burn. Following a public meeting, the Trust enlisted the assistance of the University scientists who introduced the idea of creating a wetland to treat the minewater discharges. After extensive lobbying and funding appeals, the Trust succeeded in gaining support from the National Rivers Authority for the construction of a pilot wetland area which was to be built by the University scientists.

The third interested party was an arts organisation, Sunderland-based Artists' Agency, which was invited by the Trust and the University to coordinate the fundraising for the creation of the permanent Wetland. This was no usual artists' agency, however, for it was already engaged in research into ways in which artists could contribute to what might at first sight seem to be an unlikely 'artistic' subject, the problem of water pollution and potential solutions to it.

The creation of wetlands or reedbeds has been successfully used to treat pollution from landfill sites in other parts of the UK and Europe. They are popular because they are low-cost, require no chemicals and may also serve as a habitat for wildlife. In this case excavation of the site had to be kept to a minimum because, as it had formerly been used as a finings pond for the local pit, it was highly contaminated with aluminium and iron. The solution was to create a compost wetland using waste materials from the coal-fired power station industry with a combination of compost made from cattle and horse manure and household waste, and limestone. Embankments around the Wetland were constructed on two levels, forming two distinct cells, one with two islands used to assist in flow distribution. The action of bacteria in the compost, together with the very slow movement of the water, would reduce the acidity and remove the metal contaminants, turning them from a dissolved form into a solid form.

Artists' Agency commissioned artist Helen Smith to work with the village Trust and the scientists. Under the supervision of civil engineer Professor John Knapton, she helped design a jetty to enable people to come close to the cleansing process. A glass feature was incorporated into the bridge allowing visitors to observe the polluted water flow into the Wetland, and the return of the cleansed water to the Stanley Burn was to be observed from the existing footpath. The Trust members constructed steps and a kissing gate in order to deter access by vehicles, and botanist Neville Hart planted up parts of the periphery of the Wetland and the access route, with the involvement of members of the Trust and local youth groups.

Artists' Agency and Helen Smith were not simply consultant decorators. Helen Smith had to understand the technical details, the chemical analyses, the mathematics of the water flow and the methods of site construction. Her most important role, however, was ensuring access to the site – attractive and effective physical access certainly – but also an access of understanding, so she was not simply creating artworks but contributing to the intrinsic purpose of the site. The artworks are therefore not set apart but are genuinely functional. Besides helping to create the landscape features in the heart of the Wetland, Helen Smith designed solar-powered monitoring stations along the site, incorporating sound-broadcast radio and digital sound-editing, to inform the visitor about the environmental monitoring processes and to reflect the personal observations of the local inhabitants. Artists' Agency also commissioned craftworker Lee Dalby to create willow-sculptures, after consulting scientists about the Wetland's rich geology as an inspirational source.

The scientific remedy in itself has worked and there are marked reductions in iron, aluminium and manganese concentrations and substantial increases in alkalinity. This visible physical regeneration seems to serve as a metaphor for a modest social regeneration in an area which has been blighted by unemployment and disadvantage and damaged by the ravages of its industrial past. The project as a whole is an astonishing achievement for a local pressure group. It has also made exceptional demands on the university scientists involved and on Artists' Agency, both of which groups now have a long-term commitment to the area. The scientists have a continuing obligation to monitor the water quality but they are not simply detached observers. Real friendships have been forged and a sense of ownership has emerged. Both scientists and artists share with the village community an enduring central vision which combines an application of knowledge and skill with imagination, lateral thinking, sensitivity and humour, bringing an entirely new dimension to the re-creation of community values through the addressing of real difficulties. The project has had its problems. The glass-bottomed bridge was smashed by vandals and has had to be replaced by a metal grid, and the site now needs to be watched over constantly. The work will never be finished and all the parties will have to remain committed.

Social deprivation is a continuing problem even where people make an effort to improve their lives. The older people of the village with their memory of real 'community' have been better able to sustain their commitment than the young, with their more fragmented culture. Artists, scientists and village participants have had complicated roles – as social workers, teachers, youth leaders, planners, scientists, engineers, botanists, information technologists and landscape designers. In turn all the professionals in these various areas have had to learn to develop a more holistic vision. Lucy Milton of Artists' Agency writes:

> You can't have a purely ecological project. The village, like any community, has a lot of problems and I know that the Agency couldn't begin to solve them – only, perhaps, to identify some issues, and this doesn't necessarily help. I'm sorry to give you this bleak but realistic vision – one of the things we had to learn was to limit the scope of what we could and could not deliver.

Since the community arts movement of the 1960s and 1970s a number of artists' agencies have become established and have developed an expertise in identifying problem areas and working closely with local people to find imaginative solutions. Artists have played an important role in creating visually arresting or attractive artworks to please the eye and express local pride, particularly in areas of industrial or urban blight. But the making of art represents only one fraction of a project's achievements. It is not simply a matter of designing solutions and then withdrawing. The implementation of those solutions require social skills and personal commitments. A project such as the Wetlands provides an opportunity for artists – and scientists – to join a community and to make their work relevant to people's lives.[9]

Art, science and society

The mediation between science and society requires intelligence, tolerance, humour, inventiveness and even bravery and it is encouraging to discover artists who are eager to take on this responsibility and in so doing maintain a balance between demystification and sharp questioning. The artworks which result cannot properly be appreciated without an understanding of the contexts from which they have arisen. They are therefore not simply gallery exhibits but part of a more nebulous interaction between science and the public. We are reminded again of Leavis's prescient words:

> The advance of science and technology means a human future of change so rapid and of such kinds, of tests and challenges so unprecedented, of decisions and possible non-decisions so momentous and insidious in their consequences, that mankind – that is surely clear – will need to be in full possession of its full humanity.[10]

Seen: Unseen, the Quaking Houses Wetland Project, 1998. Courtesy of University of Newcastle upon Tyne and Artists' Agency, Sunderland. Photo: Colin Cuthbert.

Notes and references

1. Jo Shapcott, from 'Delectable Creatures', in *My Life Asleep* (Oxford, Oxford University Press, 1998).
2. Julia Hinde, 'What will be the big scientific breakthroughs of the 21st century? Leading scientists peek into the future…', in *Times Higher Education Supplement*, 29 January 1999, p. 20.
3. John Horgan, 'Nothing left to learn?', in *Times Literary Supplement*, 29 January 1999, p. 31, a review of *What Remains to be Discovered* by John Maddox (London, Macmillan, 1999). Horgan's own book is *The End of Science* (London, Little, Brown, 1997).
4. Lewis Wolpert has said this on a number of occasions. He explains his thinking in *The Unnatural Nature of Science* (London, Faber and Faber, 1992).
5. Sources for the section on James Acord are as follows: James Flint, catalogue essay in *Atomic*, exhibition catalogue (London, Arts Catalyst, 1998); the Robert Weiner quotation used by James Flint is from *I am a Mathematician* (Garden City, NY, Doubleday, 1956), pp. 307–8; Kate Worsley, 'Nuclear Acord', in *Times Higher Education Supplement*, 4 September 1998; James Acord at The Eye of the Storm, the Arts Catalyst conference, 1998; *Atomic* on the Arts Catalyst website (artscat.demon.co.uk); Nicola Triscott and Rob La Frenais, Arts Catalyst. The exhibition *Atomic*, at Imperial College, London, 1998, also featured the work of two other artists concerned with the nuclear industry: Carey Young's *Legacy Systems*, which examined the original Soviet space race from the artefacts treasured or abandoned in contemporary Russian space museums or space cities and Mark Waller's *Glowboys*, a short film of a fictional British nuclear power plant.
6. Sources for the section on Cornelia Hesse-Honegger are: *The Future's Mirror*, exhibition catalogue (Newcastle upon Tyne, Locus +, 1998), 'Preface' and 'Case Studies' by Cornelia Hesse-Honegger, 'Forces of the Small: Paintings as sensuous critique' by Peter Suchin, 'The Work of Cornelia Hesse-Honegger: A scientist's perspective' by Georges B. Dussart; and *After Chernobyl: Cornelia Hesse-Honegger*, exhibition catalogue (Bern, Bundesamt für Kultur, 1992).
7. *Detail of Inheritance (Hotel)* and quotations are from *Cultivar and Inheritance (Hotel)*, exhibition catalogue (Bourn, Wysing Arts, 1999). Other information comes from discussions with the artist Neal White and Trystan Hawkins of Wysing Arts (Fox Road, Bourn, Cambridgeshire, CB3 7TX, 01954 718881, infor@wysing.demon.co.uk); also with Michael Rhodes, Barb Gorick and Denise Clark at the Human Genome Mapping Project Resource Centre, Hinxton, Cambridge.
8. Report by Paul L. Younger and Adam P. Jarvis, *The Quaking Houses Community Wetland for Minewater Remediation*, January 1998.
9. The Wetland won two major awards in 1998: a commendation for Best Practice in Community Regeneration in the British Urban Regeneration Agency (BURA) Charitable Awards, and as a category winner in the Henry Ford European Conservation Awards. Sources for the section on the Wetland are: Lucy Milton, *The Seen and Unseen*, Artists' Agency's report (Charlbury, John Carpenter Publishing, 1998); Paul Younger, 'The Gavinswelly Wetland', in *The Northern Review*, 2, Winter 1995; Artists' Agency newsletters and promotion material and conversation and correspondence with Lucy Milton of Artists' Agency (18 Norfolk Street, Sunderland, Tyne and Wear, SR1 1EA). See also John Griffiths and Penny Kemp, *Seen: Unseen* (Charlbury, John Carpenter, 1999), and the Artists' Agency conference report for Seen: Unseen – Collaborative Solutions to Pollution (November 1999).
10. F.R. Leavis, *Two Cultures? The Significance of C.P. Snow* (London, Chatto and Windus, 1962), p. 26.

Select Bibliography

This bibliography is mainly of science books suggested for further reading, and also books that examine art and science together. It does not include the many art catalogues relating to the work of artists mentioned in the text, some of which are cited in the relevant chapter endnotes.

Beate Allert, ed., *Languages of Visuality: Crossings between Science, Art, Politics and Literature* (Detroit, Wayne State University Press, 1996).

John D. Barrow, *Theories of Everything: The quest for ultimate explanation* (Oxford, Oxford University Press, 1990).

Marga Bijvoet, *Art as Enquiry: Toward new collaborations between art, science and technology* (New York, Washington DC, Baltimore, etc., Peter Lang, 1997).

Susan Blackmore, *The Meme Machine* (Oxford, Oxford University Press, 1999).

Margaret A. Boden, ed., *Dimensions of Creativity* (Cambridge, MA, MIT Press, 1994).

David Bohm, *On Creativity*, ed. Lee Nichol (London and New York, Routledge, 1998).

David Bohm, *The Special Theory of Relativity* (London, Routledge, 1996).

Dee Breger, *Journeys in Microspace: The art of the scanning electron microscope* (New York, Columbia University Press, 1994).

Louisa Buck, *Moving Targets: A user's guide to British art now* (London, Tate Gallery Publishing, 1997).

John Carey, ed., *The Faber Book of Science* (London, Faber and Faber, 1995).

Rita Carter, *Mapping the Mind* (London, Weidenfeld and Nicolson, 1998).

Peter Coveney and Roger Highfield, *Frontiers of Complexity: The search for order in a chaotic world* (London, Faber and Faber, 1995).

Paul Davies, *God and the New Physics* (London, Penguin Books, 1990).

Paul Davies and Julian Brown, eds., *Superstrings, A theory of everything?* (Cambridge, Cambridge University Press, 1992).

Richard Dawkins, *The Blind Watchmaker* (London, Penguin Books, 1986).

Richard Dawkins, *The Selfish Gene*, 2nd edn (Oxford, Oxford University Press, 1989).

Richard Dawkins, *Unweaving the Rainbow* (London, Allen Lane, The Penguin Press, 1998).

Daniel Dennett, *Darwin's Dangerous Idea* (London, Allen Lane, The Penguin Press, 1995; New York, Simon and Schuster, 1995).

David Deutsch, *The Fabric of Reality* (London, Allen Lane 1997; Penguin 1998).

Jared Diamond, *Guns, Germs and Steel* (London, Vintage, 1998).

Denis Dutton and Michael Krausz, eds., *The Concept of Creativity in Science and Art* (The Hague and Boston, MA, Kluwer Academic Publishers, 1981).

William Ewing, *Inside Information: Imaging the human body* (New York, Simon and Schuster, 1996).

Paul Feyerabend, *Against Method* (London, New Left Books, 1975).

Richard Feynman, *The Character of Physical Law* (London, Penguin, 1995).

Michel Foucault, *The Order of Things: An archaeology of the human sciences* (London and New York, Tavistock, 1970).

Michael Frayn, *Copenhagen* (London, Methuen Drama, 1998).

Peter Galison, *Image and Logic: A material culture of microphysics* (Chicago, University of Chicago Press, 1997).

J.A. Goguen and Robert K.C. Forman, eds., *Journal of Consciousness Studies: Controversies in science and the humanities* (Imprint Academia; ISSN 1355 8250). Records of the annual Tucson conferences on consciousness.

E.H. Gombrich, *The Story of Art*, 16th edn (London, Phaidon Press, 1995).

Brian Goodwin, *How the Leopard Changed its Spots* (London, Phoenix, 1994).

Stephen Jay Gould, *Leonardo's Mountain of Clams and the Diet of Worms* (New York, Harmony

Books, 1998; London, Jonathan Cape, 1999).
Stephen R. Graibard, ed., *Art and Science* (Lanham, MD, and London, University Press of America, 1986).
Susan Greenfield, *The Human Brain: A guided tour* (London, Weidenfeld and Nicolson, 1997).
R.L. Gregory, *Concepts and Mechanisms of Perception* (London, Duckworth, 1974).
R.L. Gregory, *Eye and Brain*, revised edn (Oxford, Oxford University Press, 1997).
R.L. Gregory, *Mirrors in Mind* (London, Penguin, 1998).
R.L. Gregory, *The Oxford Companion to the Mind* (Oxford, Oxford University Press, 1987).
John Gribbin, *Companion to the Cosmos* (London, Weidenfeld and Nicolson, 1996).
John Gribbin, *Q is for Quantum: Particle physics from A to Z* (London, Weidenfeld and Nicolson, 1998).
Craig Harris, ed., *The Leonardo Almanac: International resources in art, science and technology* (Cambridge, MA, MIT Press, 1993).
Tony Hey and Patrick Walters, *Einstein's Mirror* (Cambridge, Cambridge University Press, 1997).
Mae-Wan Ho, *Genetic Engineering – Dream or nightmare?* (Bath, Gateway Books, 1998).
John Horgan, *The End of Science* (London, Little, Brown, 1997).
Robert Hughes, *The Shock of the New* (London, Thames and Hudson, 1980, 1996).
Charles Jencks, *Modern Movements in Architecture*, 2nd edn (London, Penguin Books, 1985).
Caroline A. Jones, Peter Galison, with Amy Slaton, eds., *Picturing Science, Producing Art* (New York, Routledge, 1998).
Steve Jones, *The Language of the Genes* (London, HarperCollins, 1993).
Martin Kemp, *From a Different Perspective: Visual angles on art and science from the Renaissance to now* (New York, Basic Books, 2000).
Martin Kemp, *The Science of Art: Optical themes in Western art from Brunelleschi to Seurat* (New Haven, Yale University Press, 1990).
Martin Kemp, *Structural Intuitions: The 'Nature' book of art and science* (Oxford, Oxford University Press, 2000).
György Kepes, *The New Landscape in Art and Science* (Paul Thobal & Co., 1956).
Arthur Koestler, *Insight and Outlook: An enquiry into the common foundations of science, art and social ethics* (New York, Macmillan, 1949).
Thomas Kuhn, *The Structure of Scientific Revolutions*, 2nd edn (Chicago, University of Chicago Press, 1970).
Bruno Latour, *Pandora's Hope: Essays on the reality of science studies* (London, Harvard University Press, 1999).
Bruno Latour, *Science in Action: How to follow scientists and engineers through society* (Milton Keynes, Open University Press, 1987).
Bruno Latour and Steve Woolgar, *Laboratory Life: The construction of scientific facts* (Princeton, Princeton University Press, 1979).
F.R. Leavis, with an essay by Michael Yudkin, *Two Cultures? The Significance of C.P. Snow* (London, Chatto and Windus, 1962).
David Lodge, *Modern Criticism and Theory* (London and New York, Longman, 1988).
John Maynard Smith, *Shaping Life: Genes, embryos and evolution*, in the series *Darwinism Today* (London, Weidenfeld and Nicolson, 1999).
Duncan McCorquodale, Naomi Siderfin and Julian Stallabrass, eds., *Occupational Hazard: Critical writing on recent British art* (London, Black Dog Publishing, 1998).
Arthur I. Miller, *Insights of Genius: Imagery and creativity in science and art* (New York, Copernicus, 1996).
Steven Mithen, *The Prehistory of the Mind* (London, Thames and Hudson, 1996).
National Advisory Committee on Creative and Cultural Education, *All Our Futures: Creativity, culture and education.* Report to the Secretary of State for Education and Employment and the Secretary of State for Culture, Media and Sport (Sudbury, Suffolk, DfEE Publications, Crown Copyright, 1999).
Steven Pinker, *How the Mind Works* (London, Allen Lane, The Penguin Press, 1997).
Karl R. Popper, *The Logic of Scientific Discovery*, trans. by the author with Julius Freed and Lan Freed (London, Hutchinson, 1959).
V.S. Ramachandran and Sandra Blakeslee, *Phantoms in the Brain* (London, Fourth Estate, 1998).
Harry Redner, *A New Science of Representation: Towards an integrated theory of representation in science, politics and art* (Boulder, CO, Westview Press, 1994).
Report of the International Symposium on the Natural Sciences and the Arts: Aspects of interaction from the Renaissance to the 20th century, Uppsala, S. Academiae Ubsaliensis (Stockholm, Almqvist and Wiksell International, 1985).

Matt Ridley, *The Origins of Virtue* (London, Viking, 1996).
Steven Rose, *Lifelines: Biology, freedom, determinism* (London, Allen Lane, The Penguin Press, 1997).
Steven Rose, ed., *From Brains to Consciousness* (London, Allen Lane, The Penguin Press, 1998).
Steven Rose, R.C. Lewontin and Leon J. Kamin, *Not in our Genes* (London, Penguin Books, 1984).
Albert Rothenberg, *The Emerging Goddess: The creative process in art, science and other fields* (Chicago, University of Chicago Press, 1979).
Oliver Sachs, *The Man who Mistook his Wife for a Hat* (London, Gerald Duckworth, 1985).
Raman Selden, *The Theory of Criticism* (London, Longman, 1988).
Lee Smolin, *The Life of the Cosmos* (London, Weidenfield and Nicolson, 1997).
C.P. Snow, *The Two Cultures, with an introduction by Stefan Collini* (Cambridge, Cambridge University Press, 1993).
Alan D. Sokal and Jean Bricmont, *Intellectual Impostures* (London, Profile Books, 1998).
Robert L. Solso, *Cognition and the Visual Arts*, Cognitive Psychology Series (Cambridge, MA, MIT Press, 1996).
Christa Sommerer and Laurent Mignonneau, eds., *Art@science* (Vienna and New York, Springer-Verlag, 1997).
Tom Sorell, *Scientism* (London and New York, Routledge, 1994).
Cyril Stanley Smith, *From Art to Science: Seventy-two objects illustrating the nature of discovery* (Cambridge, MA, MIT Press, 1980).
Ann Thomas, ed., *Beauty of Another Order: Photography in science* (New Haven and London, Yale University Press, in association with National Gallery of Canada, Ottawa, 1997).
James D. Watson, *The Double Helix*, ed. Gunther S. Stent (London, Weidenfeld and Nicolson, 1981).
A.N. Wilson, *God's Funeral* (London, John Murray, 1999).
Lewis Wolpert, *The Unnatural Nature of Science* (London, Faber and Faber, 1992).
Duncan Wu, ed., *Companion to Romanticism* (Oxford, Blackwell, 1998).
Semir Zeki, *Inner Vision: An exploration of art and the brain* (Oxford, Oxford University Press, 1999).

The Authors

Ken Arnold is Exhibitions Manager at the Wellcome Trust, overseeing two galleries dedicated to exhibitions dealing with the culture of medicine (its science, art and history) and project-managing a number of initiatives that bring together science and art, most notably the funding of Sciart. He also writes and lectures on the culture of museums past and present. Before his arrival at the Wellcome Trust in 1992, he worked for various museums on both sides of the Atlantic. He studied Natural Sciences at Cambridge University and his doctorate at Princeton University was on the history of English museums.

Richard Bright is the founder and Director of the Interalia Centre, Bristol, which provides an international forum for exploring the relationship between the arts and sciences. He studied Fine Art at Central School of Art and Camberwell School of Art and also has a degree in Physics from the Open University; his work as an artist draws much from scientific concepts of mathematical and natural and physical processes. He has lectured widely and participated in a number of broadcasts on art and science, and given courses on colour theory and practice, geometry and art and on scientific concepts in early twentieth-century art. He has contributed essays to exhibition catalogues for the artists Susan Derges and Andy Goldsworthy and to a book on the work of James Turrell.

Andrea Duncan is Head of Painting in the School of Art and Theory at the University of East London. As a practising artist she works with painting, text and photography and has explored themes of body-mind relationship and dysfunction in such series as *A Little Death at Nemi* (1993) and *Mileva and the Curved Field* (1996). Her research field includes the role of metaphor within art and literature and the specialised languages of science. She is involved in the Humanscape project with the Department of Haematological Medicine at King's College Hospital and has published work in the areas of psychoanalysis and analytical psychology, visual theory and contemporary practice.

Siân Ede is Arts Director at the Calouste Gulbenkian Foundation where she runs the Arts grant-giving programme and commissions publications to promote new initiatives (professional and non-professional) in the arts in Great Britain and the Republic of Ireland. She has worked in theatre, artist education and journalism, was Drama Officer at the Arts Council of Great Britain and taught the Arts and Education course at the Department of Arts Policy and Management, City University, London. She has written, reviewed and spoken at conferences on the arts and arts education.

Martin Kemp is Professor of the History of Art at the University of Oxford, and was British Academy Wolfson Research Professor from 1993 to 1998. He trained in Natural Sciences and Art History at Cambridge University and has spent most of his career in Scotland (Universities of Glasgow and St Andrews). He is the author of *Leonardo da Vinci: The marvellous works of nature and man* (1981, winner of the Mitchell Prize); *The Science of Art: Optical themes in Western art from Brunelleschi to Seurat* (1990); *Behind the Picture* (1997) and *Imagine e Verità* (1999). He is currently writing a regular column in *Nature* on art, science, image and culture. Work in press includes *Structural Intuitions: The 'Nature' book of art and science* and *From a Different Perspective: Visual angles on art and science from the Renaissance to now.*

Mike Page is based at the Medical Research Council Cognition and Brain Sciences Unit, Cambridge University (formerly the Applied Psychology Unit). His research interests include the connectionist modelling of human memory, with particular emphasis on memory for serial order; he has recently contributed an essay on 'Connectionist Modelling in Psychology' to *Behavioral and Brain Sciences* (in press). He read Engineering Science at the University of Oxford before completing a doctorate at the University of Wales, Cardiff, on 'Modelling Aspects of Music Perception Using Self-organising Neural Networks'. He has also spent a year at the Department of Psychology, St George's Hospital Medical School.

Deborah Schultz is a lecturer at Central St Martin's College of Art and Design and a Research Fellow at the Henry Moore Institute, Leeds. She has completed a doctorate at the University of Oxford on 'Marcel Broodthaers: Strategy and Dialogue', and has recently contributed an essay on 'Conceptual Play: Artists' books by Marcel Broodthaers and Maurizio Nannucci' to *A Spectre at the Feast: Concrete poetry, artists' books and conceptual art* (Seattle, Clarté Press, 2000).

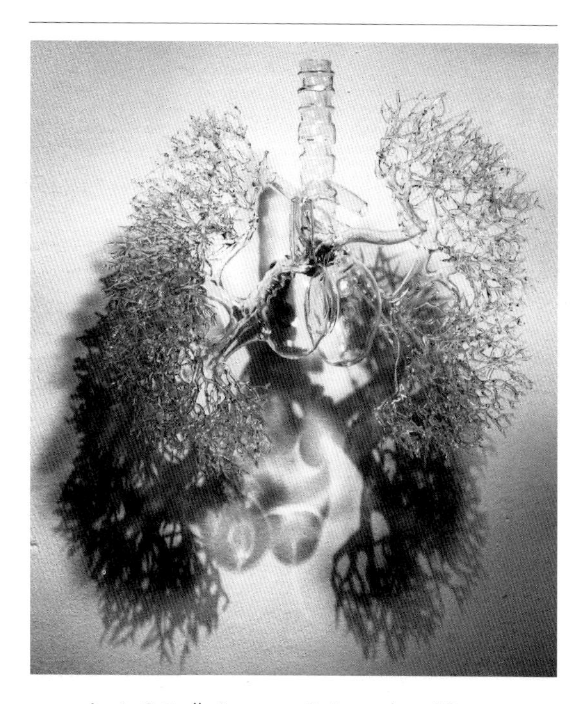

Annie Cattrell, *Access*, 1998. Pyrex glass, life-size heart and lungs. Courtesy of the artist. Photo: Peter Cattrell.

Acknowledgements

We would like to thank the following people:

The Steering Group
Marjorie Allthorpe-Guyton, Visual Arts Department, Arts Council of England
Ken Arnold, The Wellcome Trust, London
Frances M. Ashcroft, University Laboratory of Physiology, University of Oxford
Paul Bonaventura, The Laboratory at the Ruskin School of Drawing and Fine Art, University of Oxford
Richard Bright, The Interalia Centre, Bristol
Daniel Brine, Combined Arts Department, Arts Council of England
A.S. Byatt, writer
Andrea Duncan, School of Art and Theory, University of East London
Stephen Farthing, artist, Ruskin Master of Drawing, University of Oxford
Chris Isham, Theoretical Physics Group, Imperial College of Science, Technology and Medicine, University of London
Gillian Thomas, At-Bristol (Chair)
Nicola Triscott, The Arts Catalyst, London

and
James Acord, artist
Beverley Ager, King's College London
Heather Ackroyd and Dan Harvey, artists
Miriam Amari, Lisson Gallery, London
Robert Andrews, Royal Society of Chemistry, Cambridge
Bergit Arends, National Institute for Medical Research, London; Sciart Consortium
Michael Benson, The London Institute
Cathy Bereznicki, National Endowment for Science, Technology and the Arts
Therese Bergne, curator
John Bewley, Locus+, Newcastle upon Tyne
Tim Bliss, Division of Neurophysiology, National Institute for Medical Research, London
Mario Borza, water sculptor
Jill Bottomly, *New Scientist*
Jack Bradley, Royal National Theatre, London
Tim Britton, Forkbeard Fantasy, Bristol
David Brodie, Spacetime Projects, Derby
Sonya Brown, The Wellcome Trust Medical Photographic Library, London
Tim Brown, Poetry Society, London
Anastasia Calder, The Arts Catalyst, London
Gabby Campbell, Natural History Museum, London
Annie Cattrell, artist
The Helen Chadwick Estate
Dinos and Jake Chapman, artists
Donna Charielle, Astrographics Publishing
Peter Chatwin, artist
Denise Clark, Human Genome Mapping Project Resource Centre, Hinxton, Cambridge
Michael J. Clark, The Wellcome Trust, London
Valerie Connor, Project Arts Centre, Dublin
Linda Copperwheat, Saatchi Gallery, London
Dorothy Cross, artist
Marianne Deconinck, Science Museum, London
Susan Derges, artist
Simon Dickson, Royal Botanic Gardens, Kew
Helen Dobson, Corbis
Philip Dodd, Institute of Contemporary Arts, London
Julie Dorrington, Wellcome Medical Photographic Library
Kitsou Dubois, artist
Anthony Everitt, arts consultant
José Alberto Feijó, Faculty of Science, University of Lisbon
Bronac Ferran, Arts Council of England
Jemima Fishburn, Institute of Contemporary Arts, London
Martha Fleming, artist
Felice Frankel, artist
Martin Freeth, National Endowment for Science, Technology and the Arts
Rhona Garvin, Science Museum, London
Stefan Gec, artist

Bruce Gilchrist, artist
Lucy Glendinning, artist
Andy Goldsworthy, artist
Barb Gorick, Human Genome Mapping Project Resource Centre, Hinxton, Cambridge
Antony Gormley, artist
Mike Greenhaugh, Department of Physics and Astronomy, University of Wales, Cardiff
Richard Gregory, Department of Experimental Psychology, University of Bristol
Jennifer Greitschus, Visual Arts Department, Royal Festival Hall, London
Sergio Gulbenkian, Instituto Gulbenkian de Ciência, Lisbon
Karen Guthrie, artist
George Hardie, Department of Graphic Design, University of Brighton
Jenny Harris, Royal National Theatre, London
Michael Harrison, Kettle's Yard, Cambridge University
Mona Hatoum, artist
Trystan Hawkins, Wysing Arts, Bourn, Cambridgeshire
Gill Hedley, Contemporary Art Society, London
Margot Heller, curator
John Herwood, Institute of Physics, London
Cornelia Hesse-Honegger, artist
Basil Hiley, Professor of Theoretical Physics, Birkbeck College, University of London
Anna Hill, artist
Anna Hill, BBC Information
Susan Hiller, artist
Damien Hirst, artist
Peter Holland, School of Interdisciplinary Studies, University of the West of England, Bristol
Matthew Holley, Department of Physiology, University of Bristol
Jeremy Hooker, poet
Michael Hue-Williams, Michael Hue-Williams Fine Art Limited, London
Tim Hunkin, artist, designer, engineer
Ian Hunt, writer and critic
Sarah Hunt, Science Museum, London
Robert Hutchison, Southern Arts Board
William James, Sir William Dunn School of Pathology, University of Oxford
Adam Jarvis, Department of Civil Engineering, University of Newcastle upon Tyne
Jo Joelson, artist
Denna Jones, The Wellcome Trust, London
Suzanne Keene, Science Museum, London
Jude Kelly, West Yorkshire Playhouse, Leeds
Martin Kemp, Department of the History of Art, University of Oxford
Christine Kenyon Jones, King's College London
Department of Haematological Medicine, King's College Hospital, London
James Kirkup, poet
Sandy Knapp, Natural History Museum, London
Rob La Frenais, The Arts Catalyst, London
Nancy Lane, Department of Zoology, Cambridge University
John Latham, artist
William Latham, artist
Marisa Lea, The Arts Catalyst, London
Melanie Leech, Department of Culture, Media and Sport
Janna Levin, Department of Astro-Physics, University of Sussex
Liliane Lijn, artist
Alf Linney, University College, University of London
Jenni Lomax, Camden Arts Centre, London
Rose Lord, Frith Street Gallery, London
Adam Lowe, artist
Fiach Mac Conghail, Project Arts Centre, Dublin
Rohini Malik Okon, Institute of International Visual Arts, London
Sandra Macqueen, artist
Gill Marsden
Pamela Martin, artist
Guido Martini, Tate Picture Library, London
Dale McFarland, Frith Street Gallery, London
Genista McIntosh, Royal National Theatre, London
Francis McKee, Centre for Contemporary Arts, Glasgow
Ken McMullen, The London Institute
Mark Merer, artist
Lucy Milton, Artists' Agency, Sunderland
Tim Molloy, Science Museum, London
Geoff Moore, theatre and dance director
Sally Morgan, Department of Fine Arts, University of the West of England, Bristol
Lynda Morris, Norwich Gallery, Norwich
Michael Morris, Artangel, London
Jill Nelson, British Association for the Advancement of Science
John Newbiggin, Department of Culture, Media and Sport
Jeremy Newton, National Endowment for Science, Technology and the Arts
Peter Nichols, playwright
Humphrey Ocean, artist

Gisèle Ollinger-Zinge, Musées Royaux des Beaux-Arts de Belgique
Orlan, artist
Chris O'Toole, Hope Entomological Collections, Oxford University Museum of Natural History
Cornelia Parker, artist
Venita Paul, Science and Society Picture Library, London
Antonia Payne, Ruskin School of Drawing and Fine Art, University of Oxford
Fiona Pearson, Royal Scottish National Gallery of Modern Art, Edinburgh
Sharmini Pereira, Michael Hue-Williams Fine Art Limited, London
Chris Philippidis, School of Interdisciplinary Studies, University of the West of England, Bristol
Nina Pope, artist
Jane Prophet, artist
Marc Quinn, artist
Peter Randal-Page, artist
Alistair Raphael, Whitechapel Gallery, London
Evan Reid, Human Genome Mapping Project Resource Centre, Hinxton, Cambridge
Heidi Reitmaier, Institute of Contemporary Arts, London
Anthony Reynolds, Anthony Reynolds Gallery, London
Michael Rhodes, Human Genome Mapping Project Resource Centre, Hinxton, Cambridge
Janine Richards, Artslab, London
Judith Richards, Artslab, London
Iain Ritchie, Sinfonia 21, London
Simon Robertshaw, artist
Murray Robertson, artist
Ken Robinson, Department of Education, University of Warwick; Chairman, National Advisory Committee on Creative and Cultural Education
Norma Rosso, Camden Arts Centre, London
Anthony Rowsell, Guy's Hospital, London
Benedict Rubra, artist
Robert Ruthven, artist
Charles Saatchi, Saatchi Gallery, London
Alison Sabedoria, musician
Richard Salmon, Richard Salmon Gallery, London
Simon Schaffer, Whipple Museum, Department of History and Philosophy of Science, Cambridge University
Jo Shapcott, poet
Yinka Shonibare, artist
Laurence Smaje, The Wellcome Trust, London
Tom Sorell, Department of Philosophy, University of Essex
Simon Starling, artist
Helen Storey, fashion designer
Liba Taub, Whipple Museum, Department of History and Philosophy of Science, Cambridge University
Natalie Taylor, artist
John Tchalenko, film-maker
Howard Thomas, Institute of Grassland and Environmental Research, Aberystwyth
Keith Thomson, Oxford University Museum of Natural History
Danny Thorogood, Institute of Grassland and Environmental Research, Aberystwyth
James Turrell, artist
Jenny Uglow, writer and publisher
School of Art and Theory, University of East London
Margareth Verbakel, Cordon Art, B.V.
Victoria Nero Gallery, London
herman de vries, artist
Jacy Wall, Taunton and Somerset NHS Trust
Mark Wallinger, artist
Fiona Way, Committee of Vice-Chancellors and Principals
Chris Webster, Tate Picture Library, London
Stephen Webster, Imperial College of Science, Technology and Medicine, University of London
Richard Wentworth, artist
Joanna Weston, Arts Council of Wales
Neal White, artist
White Cube Gallery, London
Kate Whitely, Biomedical Collection, The Wellcome Trust Medical Photographic Library, London
Louise K. Wilson, artist
Mike Winter, The British Council
Terry Winters, artist
Johanna Wistrom, Frith Street Gallery, London
Lewis Wolpert, science writer; University College London, University of London
Alexa Wright, artist
Y Touring Theatre Company, London
Paul Younger, Department of Civil Engineering, University of Newcastle upon Tyne
Zelda Cheatle Gallery, London

Art and Science Resources

There has been an exponential growth in art-science collaborations in the late 1990s. Many mainstream arts organisations have begun to commission or present new work or offer programmes of talks or educational activities related to science. Many science institutions have expressed an interest in working with artists. The following is a brief list of national resources.

British arts organisations which specialise in promoting art-science collaborations

The Arts Catalyst
The Arts Catalyst promotes a dialogue between the arts and science, organising collaborations in which artists explore in depth areas of scientific study, working with scientists in their research environments. Events and exhibitions are presented to the public in a range of arts, science and public venues, reaching a wide audience. Arts Catalyst also runs a programme of education activities and critical debates.
The Arts Catalyst, Toynbee Studios, 28 Commercial Street, London E1 6LS; 020 7375 3690; supanova@artscat.demon.co.uk; www.artscat.demon.co.uk

ARTLab
ARTLab was formed to initiate and support visual arts projects which propose to experiment with the concepts, processes and/or products of science, technology and medicine. Since 1997 it has been engaged by Imperial College of Science, Technology and Medicine, University of London, to organise an annual programme of art-science projects and presentations, in collaboration with college research groups; to advise on and support the Visiting Artist programme; and to help seek support for new initiatives.
ARTLab at Imperial College of Science, Technology and Medicine, William Penney Building, London SW7 2AZ; 020 7594 8442; jmr@icparc.ic.ac.uk

The Institute of Contemporary Arts (ICA)
The Institute of Contemporary Arts (ICA) runs an eclectic programme of art and science activities. These include Imaginaria, an annual digital arts award, and series of talks on subjects such as Science and Ethics and Cutting Edge Science, which address the effects of scientific discoveries on various aspects of contemporary life.
The ICA, The Mall, London SW1Y 5AH; admin: 020 7930 0493; box office: 020 7930 3647; info@ica.org.uk; www.illumin.co.uk/ica/

Interalia
Interalia was founded in 1990 to provide an international forum for the exchange of ideas that explore the relationship between the arts and sciences. The Interalia Centre was established in Bristol in 1996 and runs a programme of activities which includes interactive public meetings between leading international artists and scientists, 'science in art' residencies at art colleges, art-science business seminars and audio-visual recording projects.
The Interalia Centre, 6 Old School House, Britannia Road, Kingswood, Bristol BS15 2DB; 0117 947 8121/932 9045; Interalia@compuserve.com

The Laboratory at the Ruskin School of Drawing and Fine Art, University of Oxford
The Laboratory is the research wing of the Ruskin School and has become increasingly involved in art and science. Besides providing resources for academic and practical research, it facilitates artists' residencies, and presents occasional exhibitions, public art projects, symposia and publications.
The Laboratory, Ruskin School of Drawing and Fine Art, 74 High Street, Oxford OX1 4BG; 01865 276944; www.ruskin-sch.ox.ac.uk/lab

The Sciart Consortium
The Consortium was formed in 1999 to continue and develop the Sciart Scheme founded by the Wellcome Trust in 1996. Membership includes the

Arts Council of England, the British Council, the Scottish Arts Council, the Wellcome Trust and the Calouste Gulbenkian Foundation, with sponsorship from the National Endowment for Science Technology and the Arts (NESTA). The Scheme, which operates annually, provides opportunities for scientists (from all fields) and artists to form partnerships in order to research, develop and produce projects which reflect contemporary practice in each discipline.
The Sciart Co-ordinator, Sciart, The Wellcome Trust, 183 Euston Road, London NW1 2BE; 020 7611 8538; sciart@wellcome.ac.uk; www.wellcome.ac.uk; sciart.org (for regular art/science news)

Spacetime Projects
Spacetime Projects organises conferences, discussions, practical and educational projects in the UK in order to explore interpretations of new science. It involves artists in many of its projects.
Spacetime Projects, The College Business Centre, Uttoxeter New Road, Derby DE22 3W2; 01332 292943; spacetime@griffin.co.uk

Wellcome Trust Exhibitions Unit
The Exhibitions Unit helps further the Wellcome Trust's commitment to fostering a dialogue about medicine between policy-makers, academics (scientists and historians), artists, commentators, critics and the public. It uses two galleries – the Wellcome Gallery and the Two 10 Gallery – to present thematic exhibitions on medicine and the biomedical sciences and their cultural and historical contexts, often including the work of contemporary artists. The Exhibitions Unit launched the Sciart Scheme in 1996.
The Exhibitions Unit, The Wellcome Trust, 183 Euston Road, London NW1 2BE; exhibitions@wellcome.ac.uk; www.wellcome.ac.uk

International art-science organisations

CAMAC (Centre d'Art, Marnay Art Centre)
Camac is a creative meeting point for art, science and technology, a centre for interaction, experimentation and exploration. It supports artists' residencies, a research 'lab', student professional development schemes, exhibitions, publications and workshops.
Camac, 1 Grande Ruse, 10400 Marnay-sur-Seine; camac@club-internet.fr; www.camac.org

Leonardo, The International Society of the Arts, Sciences and Technology
Established in 1982, Leonardo serves the international art community by providing channels of communication for artists, scholars, technologists, scientists, educators, students, and others interested in the arts, with an emphasis on documenting the views of artists all over the world who use science and developing technologies in their work. It produces two journals, *Leonardo* and *Leonardo Music Journal*, and runs the Leonardo Awards Program.
Main Editorial Office: Leonardo, 425 Market Street, 2nd Floor, San Francisco, CA 94105, USA.
European Editorial Office: Leonardo, 8 rue Émile Dunois, 92100 Boulogne Billancourt, France.
Leonardo On-Line: mitpress.mit.edu/e-journals/Leonardo/

Organisations which provide information or activities concerned with the public understanding of science

The British Association for the Advancement of Science
The annual festival of the BAAS has been running since 1831 and provides an important and accessible means of communicating an understanding of the concepts, methods and possibilities of science and engineering in the UK. It is held at a different university campus each September and features lectures and debates by many of the world's leading scientists.
The British Association for the Advancement of Science, 23 Savile Row, London W1X 2NB; 020 7973 3052; BA.PROG.DIRECTR@MCR1.poptel.org.uk; www.britassoc.org.uk

COPUS
The Committee on the Public Understanding of Science (COPUS) was set up in the mid-1980s by the Royal Society, the Royal Institution, and the British Association for the Advancement of Science to pioneer a programme of activities aimed at promoting a better understanding of the role of science, engineering and technology in society. It runs some of its own activities and funds others. The COPUS programme consists of: the COPUS forum for debate and discussion; COPUS booklets and workshops to share best practice and research; the COPUS grants scheme to enable key and innovative

developments in public understanding; a celebration of the best in popular science writing through the annual Rhône-Poulenc Prize for the best new science book.
COPUS, c/o The Royal Society, 6 Carlton House Terrace, London SW1Y 5AG; 020 7451 2580; COPUS@royalsoc.ac.uk; www.royalsoc.ac.uk/copus

Edinburgh International Science Festival
An annual festival of science events, lectures and discussions held in the spring.
Edinburgh International Science Festival, 149 Rose Street. Edinburgh EH2 4LS; admin: 0131 530 2001; box office: 0131 220 4349 and 0131 473 2070; esf@scifest.demon.co.uk; www.edinburghfestivals.co.uk

The Royal Institution of Great Britain
The Royal Institution was founded in 1799 to promote the application of science 'to the common purposes of life' and has been fulfilling this goal ever since, through lectures and through the work of the Davy-Faraday Research Laboratory, whose members work on solid state chemistry. The Royal Institution runs a wide-ranging programme of public lectures, including the Christmas lectures for children, which are televised every year to an audience of millions, and Scientists for the New Century, a recently established lecture series which provides a showcase for young scientific talent.
Royal Institution, 21 Albemarle Street, London W1X 4BS; 020 7409 2992; ri@ri.ac.uk; www.ri.ac.uk

Artist Neal White and medical geneticist Evan Reid in discussion at the Human Genome Mapping Project Resource Centre, near Cambridge. Courtesy of Wysing Arts, Bourn. Photo: Trystan Hawkins.

The Royal Society
Founded in 1660, the Royal Society is the scientific academy of the UK, dedicated to promoting excellence in science. It has 1,300 fellows and foreign members who are elected from among the best scientists in the world. The Royal Society presents a varied programme of events throughout the year which promote, explain and discuss topical scientific issues.
The Royal Society, 6 Carlton House Terrace, London SW1Y 5AG; 020 7839 5561; www.royalsoc.ac.uk

Science museums
There are over 500 museums with science collections in the UK and some specialist science museums or science and technology centres. Most welcome artist research and an increasing number have positive arts policies, sometimes hosting artists in residence, commissioning new work, setting aside exhibition and performance spaces and exploring other opportunities to work with artists.
Information can be found on the DOMUS (Digest of Museum Statistics) database, The Museums and Galleries Commission, 16 Queen Anne's Gate, London SW1H 9AA; www.museums.gov.uk

Non-specialist science periodicals

New Scientist
Weekly, accessible, impartial, covers the latest science news; major features, pull-out information leaflets, book reviews, letters and questions page.
New Scientist, New Science Publications, Reed Business Information Ltd., 151 Wardour Street, London W1V 4BN; 020 8652 3500; www.newscientist.com

Nature
International weekly science journal publishing papers on recent research in different fields for the science community.
Nature, Porters South, 4 Crinan Street, London N1 9XW; 020 7833 4000; nature@nature.com

Index

Page numbers in *italics* refer to illustrations